Systems Biology

Systems Biology

B Madhava Puri

Bhakti Vedanta Institute
of Spiritual Culture and Science
Princeton, New Jersey, USA

Readers interested in the subject matter discussed in this book are encouraged to contact:

Krishna Keshava Das
krishna.keshava.dasa@bviscs.org
www.bviscs.org

The book cover was created with the assistance of generative AI.

First Printing, 2026

Published by the Bhakti Vedanta Institute of Spiritual Culture and Science Princeton, New Jersey, United States of America

Cataloging-in-Publication Data

Puri, Bhakti Madhava
Systems biology as the scientific understanding of life beyond reductionism: Science & scientist 2024 conference proceedings
Editorial advisor Bhakti Madhava Puri | editor Krishna Keshava Das
Includes bibliographical references
Print ISBN: 978-1-7349089-6-1
Ebook ISBN: 978-1-7349089-7-8

These conference proceedings are dedicated to Dr. B. Mādhava Purī (1943-2025), the visionary behind the Science & Scientist annual international conference series since 2013, and the Serving Director of the Bhakti Vedanta Institute of Spiritual Culture and Science in Princeton, New Jersey, USA.

"I am saddened to learn of Dr. Purī's passing. May his memory be a blessing. I am glad he was interested in Barbara McClintock. Her work was not appreciated as widely as it deserved. Dr. Purī must have been a very wise man."

:: James A Shapiro | Molecular Biologist and author of Evolution: A View from the 21st Century. Fortifed. (2022)

CONTENTS

The inadequacy of *reductionism* to describe living entities does not eclipse the importance of *reduction* as a mode of analysis in biology. While the former "is an ontological claim about reality" [1] presuming that wholes are nothing but the sum of their parts, the latter is a method of science that temporarily isolates phenomena at a given level of organization to ask specific questions and attain clearer details of the part in order to *reintegrate this information in the context of the whole.* The distinction between reductionism and reduction can be seen in nonreductionist biological disciplines like systems biology, a field that largely reaffirms the necessity and validity of the teleological (purposiveness / goal-directedness) viewpoint in biology. Systems biologists seek to "answer questions at the level to which they are most appropriate and then use that insight to probe down and up towards the other levels," [2] such that "we should ascribe functions and purposes to the level at which they make sense, which is the level at which they constrain the interactions of the system at lower levels. This constraint is also what canalizes those interactions to serve the natural purposiveness of organisms." [3] Examples of top-down biological purposes constraining lower levels of organization include (1) the role of the heart in the circulation of blood, (2) the role of kidney tubules in creating countercurrent flow, (3) the role of Darwin's gemmules ensuring continuity of communication of characteristics in the organism and to the inheritance of later generations, and (4) Hodgkins cycles, i.e. the electrical activity of cells in an organism. [4]

Systems biology described thus far considers systems as a "fundamental ontological category" [5] i.e. they are fundamentally more than the sum of their parts – a consideration based on empiric observation at higher levels of biological organization. This approach, which finds its roots in the philosophically informed thoughts of systems theorists and organicists throughout the mid-1900s, has come to be known as 'systems-theoretic biology' in contrast to 'pragmatic systems biology.' [6] While the former is characterized by top-down modeling, the latter prefers bottom-up models, as seen in a graphic from the U.S. Department of Energy's Genome to Life Program which tried to frame systems biology in this way. [7]

In the wake of the Human Genome Project, 'pragmatic systems biologists' emerged at the turn of the century to integrate new genomic data with mathematical models of living phenomena in hopes of achieving outcomes requiring computation of large data sets, like improving the predictive capacity of medicine by identifying genetic indications of potential diseases before they develop. Since they favor a reductionist bottom-up model of whole systems that embraces the ontological claim that the whole is nothing but the sum of its parts, their use of the term "system" is a bit misleading. It is precisely due to the falsity of reductionism's ontological claim that medical practices based on genetics, like polygenic risk scores, are practically useless; these scores are supposed to anticipate genetic-predispositions toward diseases yet they produce as many false predictions as correct ones. [8] The function of the genome is constrained by higher-level processes in the organism, thus it's irrational to think that analyzing the genome could provide reliable predictions about that by which it is determined. Systems biology in its truest sense is concerned with living entities as wholes, whose parts must be understood thoroughly yet in the proper context. Thus, the genome must be understood in the context of the living cell, cells must be understood in various capacities within the living organism, and living entities must be understood in both their objective and subjective aspects.

Again, reduction (distinguished from reductionism) as a mode of analysis in biology is needed. Comprehensive descriptions of the mechanical and chemical aspects of life are necessary but not sufficient to fully and soberly comprehend living cognitive phenomena. Although a uniform mechanical-chemical framework may describe nature horizontally across bodily forms of living and nonliving nature, a multiform heterogeneous framework that subsumes or sublates mechanical and chemical aspects is required for describing nature vertically, plunging into the cognitive depths that determine biological function and activity. The uniformity of nature reduces all living and nonliving phenomena to mechanistic principles – the material and efficient aspects of cause, i.e. the physical constituents of and the external agency involved in changing an object – however, this understanding only reflects an immediate acquaintance (*ordo cognoscendi*) with the natural world. Conceiving the mediated development behind what immediately appears to the senses (*ordo essendi*) requires considering the formal and final aspects of cause, i.e. principles of formation/origination and that for the sake of which an object exists. This homogeneous framework, embracing all four aspects of Aristotelian causality, allows a critical appraisal of the similarities and distinctions between life and nonlife. Thus, while analyzing living and nonliving phenomena through material and efficient causality provides a uniform view of natural objects as atomic, molecular, or chemical conglomerates being determined by external forces like gravity, taking the perspectives of formal and final causality into account shows the distinctively biogenic or abiogenic origin of a natural object and whether it acts for the sake of itself (living) or is merely an object determined/utilized by things external to it (nonliving).

A heterogeneous conceptual framework that accommodates the similarities and differences between life and nonlife, as well as life's objective and subjective aspects, proves to be a dialectical approach. The utility of dialectical thinking is seen in the historical development of how 21st-century biologists view the dynamic between form and function. The history of morphology in general – around the late 18th and early 19th century – saw a dichotomy between the formalists and functionalists, where the former were focused solely on morphological structure as the defining feature of organisms while the latter were concerned with the function shaping form. [9] This dichotomy also concerned architects. In 1908, Frank Lloyd Wright – designer of the Guggenheim Museum in New York City – explained that "[f]orm follows function – that has been misunderstood. Form and function should be one, joined in a spiritual union." [10] Significantly, Wright is known for being inspired by organic design principles allowing him to conceive of form and function dialectically as a heterogeneous unity or unity-in-diversity. Contemporary biologists have also begun thinking about living structures in this way. Proponents of embodied physiology view living structures as processes or activity, just as much as functions are, thus the rigid distinction between form and function dissolves. [11] So, despite the immediate opposition between them – between the extended body's particular properties (form) and the process that such determinations are meant to support (function) – the mediation of dialectic thinking reveals the fundamental unity that underlies form and function such that they can be viewed as distinctive moments of a singular dynamic organic activity. Dialectical thought facilitates transcendence from the rigid compartmentalized understanding that underlies reductionism.

Liberation from Neo-Darwinian reductionism is leading cutting-edge scientists to recognize the cognitive basis of evolution, where problem-solving, decision-making (including capacities to tolerate uncertainty and harness randomness), and cooperation with others play crucial roles in the purposeful evolution of living entities. [12] The volitional, cognitive, and emotional continuity observed throughout animals and humans – where "there are transitional stages among species, not large gaps; and that the differences among many animals are differences in degree rather than in kind" [13] – is comprehended as a spectrum of increasing individual autonomy (self-determined maintenance of organismic form and function through time) [14] where organisms become more independent from – "emancipat[ed] from direct influences and fluctuations [of]" [15] – the environment, as their sophistication gradually enhances reaching the height of the human form of life. These conclusions regarding the evolution of consciousness are based on the exponentially growing body of empirical and experimentally demonstrable evidence of 21st-century biology.

While these trail-blazing scientists recognize the fundamental connection between life and cognition and some of the implications this has on evolution, the actual origin of cognition remains ever-elusive. Even the notion of a cellular basis of consciousness does not explain how the objective and subjective mix with each other, even at the cellular level. Hegel's philosophy offers a robust dialectic conceptual framework to comprehend this – such that the objective is determined by the subjective and the subjective knows itself through the objective – as well as a deeper fundamental knowledge of systems as a telescoping series of syllogisms.

This year's Science & Scientist conference hopes to (1) interface Vedantic and Hegelian philosophies [both dialectical] with organismic and systems biology, (2) humbly contribute to the development of a more philosophical conception of systems that will serve the progress of 21st-century biology, and (3) stimulate interdisciplinary dialogue aimed at deepening mutual understanding among the speakers' valuable viewpoints. We'd also like to consider questions such as (1) what experiments can distinguish between [a] the organismic level of organization exerting top-down agency on the cellular level, and [b] the cellular level exerting bottom-up agency on the organismic level? and (2) how does conceiving living entities as irreducible cognitive systems influence evolutionary theory? All forthcoming material was approved for publication by the respective conference participants. For video recordings of their presentations and the interdisciplinary dialogue, please visit **www.bviscs.org/ss2024**.

References

1. Rosslenbroich, Bernd. 2023. *Properties of Life: Toward a Theory of Organismic Biology*. The MIT Press. p 43.

2. Noble, Denis. 2017. *Dance to the Tune of Life: Biological Relativity*. Cambridge University Press. p 47.

3. Ibid. p 250.

4. Noble, Denis. "Purposive Explanations Are More Useful In Explaining Lower-Level Processes In Living Systems Than The Other Way Round." Science & Scientist 2023 Conference Presentation. https://bviscs.org/wp-content/uploads/2023/12/Denis-Noble-SS23-slides.pdf

5. Rosslenbroich. *Properties of Life*. p 98.

6. O'Malley, Maureen and John Dupre. 2005. "Fundamental issues in systems biology." *BioEssays* 27. pp 1270-1276.

7. Johnson, Gary and Marvin Frazier. 2003. "Genomes to Life: Biological Solutions for Energy Challenges." U.S. Department of Energy. https://web.archive.org/web/20160103073858/https://doegenomestolife.org/pubs/2003abstracts/GTL2003Booklet.pdf

8. Hingorani, Aroon D., et al. 2023. "Performance of polygenic risk scores in screening, prediction, and risk stratification: secondary analysis of data in the Polygenic Score Catalog." *BMJ Medicine*. p 1.

9. Ulett, M. A. 2010. "Form and Function (1916), by Edward Stuart Russell." Embryo Project Encyclopedia. https://embryo.asu.edu/pages/form-and-function-1916-edward-stuart-russell

10. "Form Follows Function." n.d. Solomon R. Guggenheim Museum. https://www.guggenheim.org/teaching-materials/the-ar-

chitecture-of-the-solomon-r-guggenheim-museum/
form-follows-function

11. Turner, J. S. 2007. *The Tinkerer's Accomplice: How Design Emerges from Life Itself*. Harvard University Press. p 20.

12. Miller, William, František Baluška, et al. 2024. "Biology in the 21st Century: Natural Selection is Cognitive Selection." *Prog Biophys Mol Biol*. p 4.

13. Bekoff, Marc. 2000. "Animal Emotions: Exploring Passionate Natures: Current interdisciplinary research provides compelling evidence that many animals experience such emotions as joy, fear, love, despair, and grief—we are not alone." *Bioscience* 50 (10): 861-870.

14. Rosslenbroich, Bernd. 2014. *On the Origin of Autonomy: A New Look at the Major Transitions in Evolution*. Springer.

15. Rosslenbroich, Bernd, Susanna Kümmell, and Benjamin Bembé. 2022. "Features of Autonomy in Human Evolution." Biological Purpose Project. https://www.biologicalpurpose.org/blog/features-autonomy-human-evolution

Michael Levin

Self-Constructing Bodies, Collective Minds: The Intersection of Computer Science, Cognitive Biology & Philosophy

Michael Levin is the Vannevar Bush Distinguished Professor of Biology at Tufts University, an associate faculty at Harvard's Wyss Institute, and the director of the Allen Discovery Center at Tufts. He has published over 400 peer-reviewed publications across developmental biology, computer science, and philosophy of mind. His group works to understand information processing and problem-solving across scales, in a range of naturally evolved, synthetically engineered, and hybrid living systems. Dr. Levin's work spans from fundamental conceptual frameworks to applications in birth defects, regeneration, and cancer.

Dr. Levin's presentation summarized his lab's cutting-edge experimental work and shared how conclusions drawn from it challenge common scientific presumptions while pushing biomedical innovation forward. His interdisciplinary work sits at the intersection of computer science, cognitive biology, and philosophy, and is revolutionizing progress in regenerative medicine and cancer research. The presentation started and concluded with the same take-away message; categorical distinctions between matter/life and machine/organism hinder experimental work and don't accurately represent unbiasedly observed phenomena. Assessing whether or not the Levin Lab's experimental work – to the extent that it was presented at this conference – justifies such a claim, will be the task of this Summary and Critique on the presentation. Significantly, Dr. Levin also places the human at the center of a dynamic continuum, with a vertical axis of embryonic and evolutionary development and a horizontal axis denoting varying technological enhancements. The rationale behind his challenging categorical distinctions stems from anticipating ethical problems in future interactions with hybrid creatures that are already starting to emerge. The framework goal was described as "Recognize, create, and relate to truly diverse intelligences regardless of composition or origin story. Move experimental work forward – new biomedical and synmorpho capabilities – and better ethics."

Summary

PART 1:

Expanding biological frameworks was the focus of the first part of the presentation. It referred to (1) experiments on Planarian flatworms proving that information storage exists outside of the brain, in addition to (2) natural and experimental examples proving the flexibility of biology (2a) developing in novel ways divergent from typical form and (2b) immediately adjusting to successful utilization of the new form. This raises the question of the non-neural source of information in a body, as well as demonstrating the plasticity of biological systems. Before transitioning to the second part of the presentation, Dr. Levin explained why plasticity exists during development, claimed that all humans start as an oocyte (immature egg cell), emphasized the need for scientific explanations focusing on scaling and transformation rather than divisive static categories, considered individual-collective dynamic, and described the various kinds of problem spaces relevant to cognitive biology.

In the Levin Lab's experimental work, scientists exploit the Planarian flatworm's dual capacity for learning and regeneration to prove that the nature of biological information is not exclusively neural. The lab trains the Planarian – which has similar neurotransmitters to humans – to get food in a particular way, then they sever its head, wait eight days for a new head and brain to grow, and observe the worm resuming getting food in the way that it was previously trained. This demonstrates that, while the brain is necessary for motor function such that during the eight days when the Planarian had no head/brain it didn't move, the information determining what behavioral signals the brain deploys is not limited to being stored in the brain. Further, upon regenerating, there's a process by which old information is imprinted onto the new brain, showing that information can move through living tissue. Dr. Levin suggested that learning more about how information is imprinted onto the brain has implications for people with degenerative brain disease via regenerating new tissue using stem cells and imprinting old memories. These observations leave us wondering, from where does the information for anatomical structure and development come?

Dr. Levin invoked an interesting phenomenon called the "Hedgehog Gall" to dismiss the common assumption that structural information comes from the genome. The Hedgehog Gall is a structure that can develop on the leaves of oak trees stung by a parasitic wasp. The gall possesses the same exact genetic information as the regular oak leaf. Yet, its structure is drastically different: rather than flat, green, and smooth, it is round, yellow/red, and spiky. The change results from the wasp "hacking" the default information of the oak leaf by injecting chemical cues containing instructions for the new structure. Dr. Levin emphasized the importance of noting that the wasp doesn't make the gall, it merely leaves instructions for the oak cells to make the gall utilizing their existing competencies. While this shows that structural information doesn't originate from genetics, it doesn't give us a positive answer to its origin.

Biological systems, as they naturally exist without human intervention, are not restricted to predictable default structures. Further, when exposed to human intervention within a lab setting, modifying an organism's structure in a manner that wouldn't occur in nature proves the sophistication of their ability to suddenly adapt to drastic change (in a matter of days) in a manner independent of macroevolution (which requires millions of years). This further demonstrates the malleability of biological systems. The Levin Lab modified a tadpole embryo by taking a cell that would become an eye and inserting it on the tail. As embryonic development proceeded with this modification, a perfect eye formed on the tail, with an optic nerve connecting to the spinal cord or gut rather than the brain. Even more amazing is that when the scientists prevent the normal eyes from forming, the tadpole learns how to see and navigate its environment using the tail eye. In Dr. Levin's words, "This demonstrates a tremendous amount of plasticity in problem-solving competency." Further, he said that "evolution doesn't make specific fixed solutions to fixed problems, but rather evolution makes problem-solving agents" such that embryos can adapt to unexpected problems and still produce a viable organism. Dr. Levin explained that plasticity exists during development because it is a process of creative reconstruction. A newly formed embryo inherits encoded instructions from its parents on how to develop, but these encoded instructions are condensed general guidelines lacking contextual details on navigating particular circumstances. He said, "The encoding process can be deductive and algorithmic, but the decoding process has to be creative because you're missing a lot of that information. You have to take that rule and say, 'but how does it apply now?' and 'what am I going to do next?'"

Wrapping up the first part of his talk, Dr. Levin emphasized the need for scientific explanations focusing on the scaling and transformation of observed cognitive capabilities – from proto-cognition to metacognition – rather than focusing on what he sees as divisive static categories such as matter/life and machine/organism. In this context, he claimed that all humans start as an oocyte (immature egg cell), which is "just physics," and develop into an individual fully endowed with human capabilities. ***In the Critique that follows this Summary, the authors will push back on the claim that "all humans start as an oocyte."*** Reinforcing the notion that non-living entities possess a rudimentary kind of cognition (proto-cognition) which is on the low end of a spectrum including high-level human self-awareness (metacognition), Dr. Levin explained that Gene Regulatory Networks (non-living chemical entities) have six different kinds of memory, including Pavlovian conditioning (trained involuntary response to a stimulus). This has implications for biomedical applications. Transitioning to the second part of the talk, Dr. Levin denied that we are a "true unified intelligence" based on observing the underlying diversity of cells permeating all aspects of the body. ***In the Critique, the authors will argue that the reality of diversity does not preclude the truth of unity.*** He also described the various kinds of problem spaces relevant to cognitive biology, such as:

- Morphospace (morphological space): all potential forms of bodily structures - "the space of all possible large-scale shapes that a group of cells can build"
- Transcriptional space: all potential varieties of gene expression
- Physiological space: all potential functions of bodily structures
- 3D behavioral space: all potential actions of an entity within an external environment

This sets up further investigation of how cellular collectives interact and pursue common goals on varying scales, which is relevant to the unanswered question of the bodily origin of structural information.

PART 2:

The second part of the presentation focused on the role of bioelectricity as the communication medium for the collective intelligence of cells. For Dr. Levin, intelligence does not mean high-level metacognition or awareness of the goals being pursued. Rather, he defines intelligence in the same way as William James (1842-1910): "the ability to get to the same goal by different means." This ability is the foundation for plasticity in biological systems, as exemplified by how newt kidney tubules accommodate the same functional goal despite being composed of experimentally modified cells. Here, the original tubule structure is composed of eight to ten cells. Then, scientists force extra genetic material into tubule cells in early development, increasing the size of the nucleus, which in turn makes the individual cells bigger to accommodate the larger nuclei. This process doesn't negatively impact the development of a normal newt, which is surprising since (1) there is an excess of chromosomes and (2) the cells are larger, but the newt remains a normal size. The cells were intelligent enough to embrace their new experimentally enlarged size and immediately make the same structure needed for a normal newt. Dr. Levin emphasizes that it did not take millions of years of evolution for the cells to adapt to this novel circumstance, but rather it was a matter of days. This experiment can be taken even further, by making the cell so big (due to accommodating increased genetic material) that only one cell creates the kidney tubule structure. The significance is that this utilizes top-down causation due to the "anatomical outcome" influencing the cell's behavior (a cell is a lower level of organization than the tubule). Further, this single-cell tubule incorporates a molecular mechanism which differs from the multi-cell tubule examples. The former utilized cytoskeletal bending while the latter used cell-to-cell communication, thus exemplifying James' definition of intelligence where the same goal is reached through different means.

Another example of intelligent problem-solving is the Levin Lab's Picasso tadpoles, where tadpole faces are artificially rearranged before turning into a frog to test if the face remains scrambled post-metamorphosis. It turns out that the developing frog can error-correct to reach the same goal (a normal frog face), even when faced with a challenge that it's never been exposed to in nature. The significance is that there's no information from its evolutionary history preparing it for this problem, yet it immediately finds a solution. This inspires questions such as "How do the cells know when the proper pattern has been achieved?" and "How does the collective consistently coordinate the outcome?" Questions regarding cellular coordination become more intriguing when achieving a desired goal requires information from cellular collectives in very different bodily locations, exemplified by a lab rat pressing a lever to get a tasty snack. The cells at the bottom of the feet press the lever while the gut cells get the treat, and no single cell both presses the lever and gets the treat, yet the actions are coordinated. Dr. Levin posed the question "When no cell has both experiences, who owns the associative memory? Collectives know things that their parts do not know." This set him up for elaborating on the communication medium for the collective intelligence of cells; bioelectricity.

Working with bioelectricity seems to be one of the main projects of the Levin Lab. The nuances of the various experiments he described up until this point in the presentation set the groundwork and context for delving deeper into this unfamiliar scientific domain, so we'll briefly reiterate the experiments and their relevance. The regenerative Planarian flatworms could be trained, and severing their head and witnessing the execution of learned behavior after regrowth demonstrated that information is stored in places other than the brain. The Hedgehog Gall forming on oak leaves stung by parasitic wasps showed that cells can make structures greatly deviating from their standard morphology, thus indicating that structural information doesn't originate in the genome since the leaf and gall have the same genetics. Modifying tadpole embryos to develop usable tail eyes showed the quick, adaptable flexibility of biology. The newt kidney tubule experiment demonstrated that cells are intelligent in that they can achieve the same goal through various strategic processes, where such strategies can be spontaneously novel rather than evolutionarily ancient. This status of intelligence was reinforced by the Picasso face tadpoles rearranging their artificially scrambled faces during metamorphosis to form normal frogs. Finally, the rat-lever-treat experiment shows that different groups of cellular collectives can coordinate activities despite having very different exclusive experiences. So, storage of learned information isn't confined to neurons; structural information doesn't derive from genetics; distinct cellular collectives somehow share exclusive information; biological systems almost instantly adapt to utilizing unusual organs in unexpected locations and course-correct during perturbations to embryonic development, demonstrating plasticity in achieving predetermined goal states.

As will be seen, the Levin Lab's work on bioelectricity demonstrates that learned and structural information, including blueprints of default morphological structures, are stored and distributed throughout the bioelectrical web that connects various cellular collectives. Scientists interfacing with this bioelectric communication network directly influence cellular development, such as changing morphology by editing cellular memories of morphological blueprints. Dr. Levin's approach to bioelectricity takes the premise of neuroscience and extends it to developmental biology. Neuroscientists measure and record electrical patterns of brain activity during information processing, try to read and decode the patterns, and make predictions based upon their study of neural activity. Here, cells in the nervous system are connected by electrical synapses and ion channel proteins on each cell's surface which "allow the cell to create an action potential and keep a particular electrical state which may or may not get propagated to its neighbors," as Dr. Levin explained. It turns out that most cells throughout the body possess ion channels, have electrical connections to neighbors, and form electrical networks, which is why the Levin Lab can apply the principles of neuroscience to developmental biology. He wants to "frame morphogenetic problem solving as intelligent behavior in anatomical space the way that neuroscientists study behavior in three-dimensional space."

Dr. Levin's group seem to be the trail blazers of studying bioelectricity, thus they've developed the first molecular tools to read and write bioelectrical states in non-neural cells, like using "voltage reporting fluorescent dye," in order to make time-lapse videos of bioelectrical activity. These videos show electrical patterns of where certain anatomy will develop before gene expression or any chemical activity takes place to start that process. Thus, Dr. Levin calls these bioelectrical patterns "informational prepatterns" and "scaffolding," while referring to bioelectricity itself as "cognitive glue" that connects individual cells to organ-level goals, as well as to other individuals. Dispersed throughout his talk, he would share insights into general experimental principles and approaches utilized by the Levin Lab. At this point, he explained that observing how biological systems naturally work is not enough for comprehensive understanding, further emphasizing the need for "perturbative experiments" which test cognitive capacities such as if or how they correct themselves.

The Levin Lab has figured out how to communicate goals through bioelectricity to cellular collectives, which is already facilitating progress in regenerative and cancer medical innovation. The method for interfacing directly with the bioelectric network *does not* utilize external electrodes, fields, frequencies, waves, radiation, or magnets. Their approach exploits naturally existing competencies in the cellular collective. Rather than micromanaging the details of constructing anatomical structures, instead, they communicate large-scale goals like "build this here," and trust the inherent intelligence of the biological system to handle the execution. By injecting a particular ion channel mRNA (referred to as bioelectric cocktail), they modify the voltage of a cell to produce the desired anatomy. For example, a fully developed eye can be grown on a gut. The trail-blazing nature of this work challenges existing developmental biology textbooks. Dr. Levin admitted the role that humility plays in navigating these experiments, as oftentimes the limitation of conclusions drawn from them is reflective of the scientists' underestimation of the biological system's capacity, not a limitation of the system itself. He gave the example of textbooks citing that only the exterior of a body can develop an eye, with the help of the PAX-6 gene, but his bioelectric approach grew an eye on the gut of a tadpole. Further experimental evidence having significant implications for regenerative medicine includes regrowing the leg of a frog. Frogs don't normally regenerate missing limbs, but after a 24-hour treatment with a bioelectric cocktail, the missing leg regrew after 45 days. It is significant that this short-term treatment produced such a long-term result. Dr. Levin has co-founded a company called Morphoceuticals Inc. with Dave Kaplan to develop wearable bioelectric technology that can induce mammalian limb regeneration.

Delving into the nuances of how influencing bioelectricity impacts biological information storage, Dr. Levin returned to conclusions drawn from work with Planarian flatworms to demonstrate that bioelectric networks are readable and rewritable. Using the Levin Lab's special bioelectric dye, they observe an electrical pre-pattern for a head on one end of the worm, and a tail on the other. After one of Dr. Levins bioelectric cocktails (ion channel mRNA injection), however, the voltage pattern of the cells changes to produce two heads and no tail. The worm still exists with the anatomy and molecular biology of a head on one end and a tail on the other, but the bioelectric pattern has been edited to hold new information of a head on both ends, which only manifests when the flatworm is injured, i.e., the second head only grows after the tail has been cut off. Dr. Levin calls this bioelectric information storage "memory" because it retains that bioelectric pattern as the new default standard such that every time the animal is severed, its segments regrow two heads (remember that Planarian flatworms can be continuously cut into pieces and the pieces produce new worms). Due to this bioelectrical intervention not requiring any change whatsoever in genetic or anatomical "hardware," this approach disproves the prevailing notion that the genome or genotype determines the phenotype of an organism.

After establishing the rationale for studying bioelectricity and explaining how it works, Dr. Levin starts wrapping up Part 2 of his presentation by exploring the limits of the scope of memory and goal-setting of certain cellular collectives, which has implications for cancer research. Recalling that at the end of Part 1, Dr. Levin denied the existence of a "true unified intelligence" – where ***the authors will argue that the reality of diversity does not preclude the truth of unity in the Critique following this Summary*** – he referred to the size of an individual entity's goal-setting potential as its "cognitive light cone." The figurative size of the cognitive light cone encompasses scaling memories and goals. A single cell doesn't know how many toes that a foot is supposed to have, but the collective memory does know; the collective knows when that morphological goal is reached which is why development stops at that point. When a single cell disconnects from the bioelectric network of the collective, it forgets the large-scale goals (its cognitive light cone shrinks) and it reverts from a multicellular lifestyle to a unicellular one. This is when cancer develops. Here, Dr. Levin acknowledged that asking philosophical questions about the boundary of the self, inquiring where the self ends and environment begins, has practical benefits for cancer research. Using bioelectrical intervention can be used (1) to diagnose when cells disconnect from their neighbors before turning into a tumor, and (2) to reintegrate a cell having genetic damage with its neighbors, thus preventing a tumor from developing by allowing the damaged cell to continue cooperating with the collective despite its deficiency. For the Levin Lab, "cancer is studied as a dissociative identity disorder of the morphogenetic intelligence."

Seemingly related to the earlier point of there being no truly unified intelligence, Dr. Levin proceeds to question the validity of one embryo corresponding to one self. This doubt is supported by experimental evidence where severing an embryo into multiple pieces allows those pieces to become individual embryos unto themselves, which can give rise to triplets, twins, conjoined twins, etc. The number of individuals is not established by genes. Dr. Levin explained that "We know in computers how much processing can be done per number of transistors in your RAM, but in biology it's a much tougher question. How many individuals per cubic millimeter of cells? We don't know what the carrying capacity of the medium is."

The final topic of Part 2 regarded not limiting beings to binary categories such as living and mechanical. This topic is also where Dr. Levin's presentation originally started. He showed a graphic of a human being at the center of a dynamic continuum. The vertical axis had inward pointing arrows indicating the embryonic and evolutionary development that shaped the human, and the newly introduced horizontal axis with outward facing arrows depicting technological modifications that can be made to the human due to the plasticity of its biology. Based on the resiliency and malleability of biological systems to form coherent living entities despite a wide range of unexpected adversity, Dr. Levin discussed the integration of engineered materials at every level of organization from the cell to the ecosystem. Based on the existing binary distinction between life and machine, he introduced concerns for ethical standards of behavior when dealing with hybrid bioengineered entities in the near future.

PART 3:

The final part of Dr. Levin's presentation was made brief, due to the level of detail of the first two parts in addition to conference time constraints. The topic regarded speculatively tracing the source of mind and body patterns. The reason for pursuing such speculation derives from the Levin Lab discovering unconventional biological systems like biobots (anthrobots and xenobots), with morphology and behavior that has no evolutionary or developmental history. Anthrobots are made from living adult human tracheal epithelial cells (they have a human genome). With some prompting from scientists, anthrobots self-assemble into clusters of single cells that recommit to a multicellular lifestyle and form little blobs capable of healing human neuron wounds. Dr. Levin wants to know why anthrobots have a preference for the form that they adopt. Thus, informed by the results of his probing experiments in synthetic morphology, he is trying to map the non-physical space of all potential forms and functions. This endeavor is justified by the existing study of the space of mathematical truths from which mathematical discoveries arise. Dr. Levin speculatively questioned whether the non-physical space of all potential forms and functions is connected to Plato's world of forms. Reiterating his framework goal, he concluded by saying "We need to be fearless in understanding that for the same reason that biochemistry doesn't tell the story of the human mind, the physics and algorithms of quote unquote machines don't tell the story of artificial cognition either. [...] We have a much wider world of very unconventional intelligences that we need to ethically relate to."

Critique

We've painstakingly attempted to articulate every aspect of Dr. Levin's presentation to help share the valuable experimental work performed by his lab which benefits regenerative medicine and cancer research. We hope that the Levin Lab's work helps many people. The following Critique is written with the same attention to detail in hopes of humbly contributing to their impact.

Dr. Levin's framework goal is to "recognize, create, and relate to truly diverse intelligences regardless of composition or origin story. Move experimental work forward – new biomedical and synmorpho capabilities – and better ethics." His presentation began and concluded with emphasizing the desire to dissolve the categorical distinctions between life and non-life (organic and mechanical entities), and denied the reality of a true unified intelligence several times. It also discussed experiments showing the following:

- Storage of learned information isn't confined to neurons
- Structural information doesn't derive from genetics
- Distinct cellular collectives somehow share exclusive information
- Biological systems almost instantly adapt to utilizing unusual organs in unexpected locations
- Biological systems course-correct during perturbations to embryonic development, demonstrating plasticity in achieving predetermined goal states
- Bioelectricity is a readable and rewritable network throughout living bodies that cellular collectives use to store information and communicate large-scale goals
- Ion channel mRNA injection therapy can bioelectrically instigate regrowth of limbs in animals that don't normally regenerate
- Non-living Gene Regulatory Networks have proto-cognition, i.e. six kinds of memory including Pavlovian Conditioning

These impressive developments demonstrate the intelligence, malleability, and resilience of biology at various scales, while also pointing to the continuity of cognitive capacities among abiotic and biotic phenomena in nature. That being said, the practical utility of life/non-life distinction remains significant for human decision making even though it does not reflect the more nuanced ontological truth of nature. Additionally, the rejection of the existence of a true unified intelligence is not substantiated by the Levin Lab's experiments and seems to be a conclusion based upon bias. In what follows, we explain some of the practical value of life/non-life distinction and suggest researching sperm-oocyte communication for progressing in the ongoing discussion of life/non-life. Then, we address the reality of unified intelligence and attempt to trace what motivated Dr. Levin's preference for his particular interpretation.

ADDRESSING LIFE/NON-LIFE DISTINCTION

During the Q&A following his presentation, Dr. Levin said "These kinds of categories [living and non-living] are only useful and should be maintained to the extent that they help make new discoveries [...and] make predictions. [...] I would maintain that you think about what work you want that distinction to do. What is it doing for us?" Some of his reasons for wanting to dissolve these binary distinctions were:

1.   They're no longer useful due to the existence of hybrids and cyborgs, where machines are fused with humans
2.   He believes there's "no reason why evolution should have a monopoly on making real minds that matter in an ethical sense. [...] Why does it have to be the random poking of cosmic rays that design real minds? No, engineers can contribute to this too, and we have to be open-minded about this otherwise we are going to risk very serious ethical lapses."

Viewing nature as a process of dynamic becoming composed of irreducible relationships, rather than the static being of discrete entities, is conducive to scientific innovation and offers a more honest perspective of natural phenomena. The Levin Lab's aspiration to dissolve the categorical distinction between life and non-life is an appropriate response to a growing consensus of nature's processual existence. It's clear that biology relies on abiotic elements for sustenance. Air, sunlight, water, and minerals are assimilated with living bodies and nurture vital processes. Perhaps this sheds light on the etymological origin of "matter" deriving from the Latin "*mater*" meaning mother, who is a nurturer. While Dr. Levin's experimental work offers substantial support for viewing nature as process, his desire for erradicating the non-life/life distinction has more to do with avoiding unethical treatment of biologically-engineered or technologically-enhanced cells and organisms, as well as artificial intelligence (software or robotic). This is where we start to find Dr. Levin's interpretation problematic. Due to his premise of "recognize, create, and relate to truly diverse intelligences *regardless of composition or origin story*" (emphasis ours), Dr. Levin runs the risk of overlooking significant differences between organisms and machines. Nature as process implies that each and every stage or moment is relevant to all other moments. The origin or formative principles are significant for understanding mature stages of mechanical and organic wholes.

Those who ignore origin story – as modern scientists have done ever since René Descartes, Francis Bacon, and others, who intentionally abandoned the formal and final aspects of Aristotelian causality – are predisposed to not recognizing the meaningful differences between mechanics and organics. Nature as process means that where a process starts and ends are relevant to a complete understanding. Scientists who undervalue the formal and final aspects of Aristotelian causality enter into nature's process and interpret their entry point as the beginning of the process itself. Prioritizing the local or immediate experience in this way is what Wilfred Sellars called the "Myth of the Given," and it leads to incomplete knowledge. Embracing origin story allows us to discern between organic internal growth and mechanical external assembly, where cognition is internally guiding the former and externally imposed upon the latter.

As Dr. B Madhava Puri has described, organisms come to be through holistic growth internally motivated by cognition, where parts (organs) interdependently support each other's maturity and the stability of the organism as a whole, while the organism as a whole is the overarching integrative purpose that the manifold organs serve. No matter what stage of an organism's development scientists specialize in, from embryology to zoology, it's essential to recognize that all aspects of the organism are a result of a process which is relevant to the immediate stage being studied. Machines come to be through aggregation of externally assembled parts that lack an internal-bond with each other, such that each and every part is replaceable without destroying the integrity of the machine. The purpose of machines are externally imposed upon them, they're externally motivated by cognition, including algorithms encoded into machines by engineers. When we look at the formation of the hybrids and cyborgs that Dr. Levin is concerned with, we can ascertain their moral and ethical standing. Did they come from sperm and egg, or nuts and bolts? If they have an organic origin, it is important to ascribe greater value than machines, so that society doesn't start prioritizing technological development over personal development (if we haven't already).

AI and robots do not have the same moral and ethical standing as organisms (especially humans), due to their lack of voluntary decision making. Thus, self-driving vehicles are not legally accountable for fatal accidents. The involuntary actions of AI are dictated by algorithms created by human engineers who are voluntary actors and accountable agents. Robots are mechanical aggregates who have an inferior value to the living organisms that make them. Blurring or disolving the distinction between life/non-life in human decision making, such that human beings and AI robots are considered comparable or equal entities, confuses people about how to properly devote attention and resources. Rather than learning how to increasingly value and sustain life, we'll forget what it even is. It's more likely to misleadingly convince innocent people that they're non-different from robots, rather than actually make robots that are truly non-different from people. But this drags human society into ignorance about their self-conscious nature by convincing people that they're nothing more than programmed machines.

The great German philosopher Immanuel Kant clearly articulated the distinction between organic and mechanical wholes 235 years ago, which still holds true today. In *Critique of Judgement* (1790), he explained:

> "[A] thing exists as a natural purpose, if it is both *cause and effect of itself*. [...] For a thing to be a natural purpose in the first place it is requisite that its parts (as regards their being and their form) are only possible through their reference to the whole. [...] In such a product of nature every part not only exists *by means of* the other parts, but is thought as existing *for the sake of* the others and the whole, that is as an (organic) instrument. [...] This can never be the case with artificial instruments, but only with nature which supplies all the material for instruments (even for those of art). Only a product of such a kind can be called a *natural purpose*, and this because it is an *organized* and *self-organizing being*. [...] In a watch one part is the instrument for moving the other parts, but the wheel is not the effective cause of the production of the others; no doubt one part is for the sake of the others, but it does not exist by their means. [...] An organized being is then not a mere machine, for that has merely *moving* power, but it possesses in itself *formative* power of a self-propagating kind which it communicates to its materials though they have it not of themselves; it organizes them, in fact, and this cannot be explained by the mere mechanical faculty of motion." [1]

As Kant described, organic wholes are simultaneously cause and effect of themselves; they produce their own content which in turn sustains them. The content (cells, tissues, organs, etc) is organized according to the particular species concept of the whole organism, while this content simultaneously maintains the individual existence of the organism. The organic whole and its parts (or participants) exist "by means of" and "for the sake of" each other; they're reciprocally cause and effect of one another. On the other hand, mechanical wholes are the effect of their parts, but the whole does not produce the parts. While a machine is assembled, its parts exist for the sake of the whole, but they weren't caused by its means. Now, Dr. Levin's Hedgehog Gall experiment demonstrated that when stung by a parasitic wasp and injected with new chemical cues (new instructions for structural organization), oak leaf cells would produce something very different from their standard morphology. At first glance, this seems to challenge the idea that the organic whole forms its participants or that they exist strictly for the whole. As it turns out, however, the gall experiment might validate the existence of the whole organism as the source of standard structural information and uniform developmental integration of organismic participants. Clearly, cells are malleable and conform to information that does not necessarily originate from their native environment. They're not strictly committed to maintaining the organism's needs. In this regard, cells even become cancerous. When forming the Hedgehog Gall, the source of information is the wasp. When cancerous, cells forget multicellular function and resort to a unicellular lifestyle. This raises the question of where an organic part's capacity to apparently act independently comes from, given it's origination in the whole. A further question is, what is the source of information when cells are properly maintaining the organism-level form and function? What are they tuned in to? The Levin Lab showed that the bioelectric network stores information and that information is shared among various cellular collectives through it, but this doesn't explain how or why that default information was originally imprinted in bioelectricity. According to Kant's explanation, the organism as a whole is the source of this ontogenetic information, while simultaneously existing as an individual due to its effect. Regarding how cellular composition is an effect of a species concept (the organismic level of organization) yet simultaneously the cells are also the cause of the individual organism, the post-Kantian German philosopher G.W.F. Hegel's conception of the "genus-process" is a more elaborate topic relevant for further consideration that won't be discussed here.

When challenged by another scientist at the conference regarding the validity of binary distinction, Dr. Levin said "Tell me where from the journey from a single oocyte – which is just a bunch of chemicals like your thermocouple [any mechanical device] – to a human, you have to tell me where the sense of experience comes in." Here, Dr. Levin is asserting that all bodies from nonliving machines to gradations of cognitive life have some degree of experience (from proto-cognition to metacognition). From our perspective, experience is understood as a subject/object relation. Dr. Levin defends his position by claiming that the fully developed cognitive experience of mature organisms is present in nascent stages in the single oocyte, which he further claims is "just physics and chemistry," and that the intensity of cognitive experience gradually scales up through embryonic development and organismic growth. Following this logic, since the oocyte is "just a bunch of chemicals" like mechanical devices, then machines must also have some degree of experience. Dr. Levin suggests that defeating his argument requires showing where cognitive experience enters into organismic development starting from the single oocyte, and if this can't be done, then experience must have been there the entire time.

In what follows, we'll offer an experimentally testable suggestion that might distinguish between mechanical matter (attenuated consciousness) and life (awakening consciousness) by considering the differences and relationship between an oocyte (unfertilized egg) and sperm (the seed of life). This is consistent with ancient Vedic wisdom describing Maha Vishnu impregnating *pradhan* (primordial matter) and triggering the unfolding of the material world, as well as intuitions of an eminent scientist from the 19th century. Microbiologist Louis Pasteur (1822-1895) held that "Life is the germ with its becoming." He explained that:

> "The mystery of life does not reside in its manifestations in adult beings, but rather and solely in the existence of the germ and of its becoming. [...] Once the germ exists, it needs only inanimate substances and proper conditions of temperature to obey the laws of its development ... it will then grow and manifest all the phenomena that we call 'vital,' but these are only physical and chemical phenomena; it is only the law of their succession which constitutes the unknown of life. [...] This is why the problem of spontaneous generation is all-absorbing, and all-important. It is the very problem of life and of its origin. To bring about spontaneous generation would be to create a germ. It would be creating life; it would be to solve the problem of its origin. It would mean to go from matter to life through conditions of environment and of matter." [2]

It seems straightforward enough that Pasteur's "germ" refers to the germ line, cells specializing in reproduction which eventually form gametes (sperm and egg cells). Germ cells are contrasted from somatic cells which make up tissues and organs. So, returning to Dr. Levin's guidelines for defeating his position as outlined above, the first step in dismantling his argument is recognizing that single oocytes don't develop into humans; only *fertilized* oocytes do. Unfertilized oocytes degrade and dispell the resources that created the egg. This was confirmed by cell biologists in 2010, as seen below. Somehow, Dr. Levin overlooked these details in his argument.

> "Sperm and oocyte fusion induces egg activation, a process that restarts the oocyte meiotic divisions, initiates embryonic development, and prevents polyspermy. Ovulated oocytes that are

not fertilized in a timely fashion degenerate, essentially wasting the energy and resources necessary to produce them. [...] The molecular machinery that mediates sperm and oocyte communication is largely unexplored, with the exception of several marine species that are amenable to biochemical analysis." [3]

Future progress in understanding the origin of cognitive experience in living entities, namely inquiring about where the subject/object relation arises, might require further research into sperm-oocyte communication. This may also shed light on the "law of succession" that Pasteur considered the "unknown of life." In 2021, reproductive scientists confirmed the ongoing debate about sperm-oocyte interplay and emphasized the "importance of the spermatozoon in fertilization and its active role in mammalian oocyte activation." [4]

THE REALITY OF DIVERSITY DOES NOT PRECLUDE THE TRUTH OF UNITY

Addressing Dr. Levin's denial of true unified intelligence, observing the underlying diversity of cells permeating every aspect of the organism as a whole doesn't detract from the whole as that for the sake of which the cells exist. Dr. Levin's presentation clearly articulated non-reductionist sentiments such as acknowledging that cellular collectives know things which none of its constituents do, such as when to stop development upon reaching a desired morphological structure. He further explained that this information of the collective is shared through the bioelectric network with individual cells who need to halt their activity. But what is the origin of bioelectricity? Is it the property of the organism as a whole? Does the bioelectric network exist in seed form within the fertilized egg, and then unfolds the instructions for development to various cells? If so, this suggests that it's a property of the whole, as we argued earlier. Dr. Levin's presentation seems to have brought out the need for delicate and nuanced dialogue regarding the similarities and distinctions between organic collectives and wholes. Collective seems to emphasize a unified plurality of micro individuals whereas whole emphasizes the singularity of a macro individual containing constituents. Many generations of cellular collectives are born and perish within the lifetime of a single organism, thus the members of the collective are constantly changing and they all must conform to the standards of the organism as a whole. Maintaining a state of wholeness throughout endless changes, i.e. sustaining an integrated purposeful order throughout development rather than succumbing to disorderly dis-integration, is the negentropic character of life. Distinguishing between hive mind collectives and genuine organic wholes is also relevant. Organic wholes aren't composed of individuals who developed independently and then integrate into hive mind collective swarms like slime molds, etc, but rather the cells, tissues, and organs are inherently integrated by developing through and for an organism's growth. The integrative purposiveness of organismic growth is proof of a truly unified intelligence, but it is a dynamic unity and not a static one.

Any given moment of the process of growth, at any level of organization, is inseparable from the totality of its ontogenesis. All particular synchronic moments imply necessary prior conditions that give rise to subsequent stages of development, and the intrinsic end or purpose that directs the totality of the diachronic process unfolding from one stage to another until maturity is achieved. The interconnectedness of synchronic moments of living process must be considered along with the diachronic wholeness in order to have a complete understanding of the unitary organism. Dr. B Madhava Puri referred to this as the Bija (seed) conception in his exchange with Dr. Levin during the interdisciplinary panel discussion at the end of the conference. Fully acknowledging the ontogenetic process of becoming facilitates perceiving the whole in its participants as the enabling context supporting their mutual interactions. Reducing an organic whole to a particular collective (again, collectives constantly change) is to fall prey to the Myth of the Given, which takes an immediate static moment of being abstracted from its ontogenesis. Since the whole is an irreducible dynamic process of becoming, neglecting to acknowledge any aspect of the process from formal to final aspects of causation (its origin to mature development), prevents one from seeing the thoroughly dynamic actuality of the living whole. Hegel critiqued the field of anatomy for reducing living process to its immediate static moments, and reifying such moments as discrete individuals.

> "In the systems constituting an embodied form *(Gestalt)* the organism is apprehended from the abstract side of lifeless physical existence: so taken, its moments are elements of a corpse and fall to be dealt with by anatomy; they do not appertain to knowledge and to the living organism. *Qua* parts of that sort they have really ceased to be, for they cease to be processes. Since the being of an organism consists essentially in universality, or reflexion into self, the being of its totality, like its moments, cannot consist in an anatomical system. The actual expression of the whole, and the externalization of its moments, are really found only as a process and a movement, running throughout the various parts of the embodied organism; and in this process what is extracted as an individual system and fixated so, appears essentially as a fluid moment. So that the reality which anatomy finds cannot be taken for its real being, but only that reality as a process, a process in which alone even the anatomical parts have a significance." [5]

Further, Hegel quoted large paragraphs of Georges Cuvier's work in order to show how the organic whole is observable in its participants, as seen in the excerpt below. Hegel acknowledged that Cuvier "could boast that from a single bone, he could learn the essential nature of the whole animal." [6] Significantly, Cuvier established the fields of comparative anatomy and paleontology. Hegel quotes from Cuvier's *Research into the Fossil Remains of Quadrupeds* (1812), where he explained that:

"'If, therefore, the viscera of an animal are so organized that they can digest only raw meat, then the jawbones, too, must be adapted for swallowing the prey, the claws for seizing and tearing flesh, the teeth for biting it off and chewing it up. In addition, the whole system of the motor organs must be adapted for pursuing and securing the prey: and the eyes, too, for seeing it at a distance. Nature must even have implanted in the brain of the animal the necessary instinct to conceal itself and to ensnare its victims. These are the general conditions of the *carnivora*; each carnivore must, without fail, combine these within itself. But particular conditions like the size, kind, and haunt of the prey, also result from particular circumstances within the general forms; so that not only the class, but also the order, the genus, and even the species, is expressed in the form of each part. 'In fact, in order that the jawbone can seize the prey, the condyle'— the organ which moves the jawbone and to which die muscles are attached — 'must have a particular shape. The temporal muscles must possess a certain bulk, and this requires a certain hollowing of the bone in which they are inserted and of the process of the cheek-bone (*arcade zygomatique*) under which they pass. This latter must also have a certain strength in order to afford sufficient support to the masticatory muscle (*masseter*).' The same principle applies throughout the entire organism: 'In order that the animal can carry away its prey, the muscles which lift the head' (the neck muscles) 'must have a certain strength; this in turn affects the form of the dorsal vertebra to which the muscles are attached, and the form of the occiput in which they are inserted. The teeth must be sharp in order that they can cut flesh and must have a solid base so that bones can be crushed. The claws must have a certain mobility'— accordingly their muscles and bones must be developed; similarly with the feet, etc.'" [7]

Moving on, upon further research for this Critique, the reason for Dr. Levin's preference for emphasizing the collective in such a way that denies true unified intelligence (a self) seems to derive from an affinity for Buddhist doctrine, specifically the concept of "no-self" and perhaps "dependent arising." The following is from a peer-reviewed paper in MDPI *Entropy* co-authored by Dr. Levin:

> "[A] Self is an illusory modelling construct created by perceptual systems of Agents. Agents construct models of causal Selves for others, and for ourselves, using the same machinery. The same mechanisms that cause an agent to act toward stress reduction in itself (even though the beneficiary of those actions is in an important sense impermanent) can be expanded to extend to other Selves. In this way, while our focus is on understanding and formulating Self in a way that is applicable to a broad range of scientific contexts, we also see ourselves as here contributing to the treatment of perennial issues in contemporary Buddhist philosophy—such as the feasibility of genuine care in a world without real individuals [...]. Similarly, with respect to the paper's main thesis regarding care as a driver for intelligence: we hope that apart from addressing contemporary scientific or social aims and practices, our discussion may as well contribute to an understanding of classic Buddhist doctrine in its own right. [...] We propose that a number of core Buddhist concepts (in particular the constructed nature of selfhood/no-self, *duḥka*, universal loving care, and the Bodhisattva) can be profitably used to challenge and to enrich the current work in diverse intelligences, including novel approaches to AI and to biology." [8]

The following excerpt regarding "dependent arising" is taken from a blog post on the Center for the Study of Apparent Selves (CSAS) website, an institution with which Dr. Levin is associated.

"In Buddhist terminology, the world of interconnected differences is referred to as 'dependent arising' (Pali: *paṭiccasamuppāda*, Sanskrit: *pratītyasamutpāda*). As a doctrine, the principle of dependent arising (DA) is typically explained with the simple statement 'when this is, that will be; when that occurs, this will occur.' Before we begin to bring this principle to bear on AI, let us here briefly notice a few of its salient features. Certain appearances notwithstanding, DA is not doctrine of natural laws that underpin a causally determined world. Rather, DA can be seen as a way of doctrinally accommodating two seemingly conflicting but equally undeniable perceptions: difference and sameness. Despite the potential for infinite differentiation that we noticed before, it is still, argues the teaching of DA, the case that, 'when this is, that will be; when that occurs, this will occur'—because a sprout, for example, can be seen to grow, quite specifically, from its seed, and a boat is rightly qualified as long when there is something else that is short in comparison. Whether we think in terms of substances or concepts, our perception of dependencies cannot be explained away and must be acknowledged as they appear. That is, arguably, the motivation behind the doctrine of DA." [9]

A theory laden affinity for "no-self" can influence interpretations of the Levin Lab's innovative experimental work. This doctrine undermines consciousness studies in biological systems by ignoring the unitary experiencer of diverse conscious experience, i.e. the center of varying degrees of willing, thinking, and feeling. The organism without a self is only a collective or totality, but to be a unity implies there must be a being-for-self such that the parts are for the whole, as previously discussed. Bhāgavat Vedānta suggests distinguishing between false-self and true-self. The numerous generations of cellular collectives that come and go through metabolic activity within the duration of an organism's single lifetime may be sufficient to conclude that identifying with the physical body produces a false-self, due to the body's transient nature.

From a more mindful assessment of our first-person experience, however, the constant changes of the body allows individuals to recognize the persistent presence which witnesses these changes, the unchanging true-self. It's this self, influenced by mundane or transcendental conditioning, that desires to adopt various worldviews such as "selfish gene," "organicism," "no-self," or "self-surrender."

In conclusion, we feel that studying Hegelian dialectics concerned with dynamic unity-in-diversity or identity-in-difference – which is complementary to Vedic wisdom's essence in Bhāgavat Vedānta – can help mediate between extremes such as mechanical-organic, materialism-idealism, monism-dualism, individualism-collectivism, and matter-spirit. Relating back to Pasteur's intuition that "life is the germ with its becoming," microbiologist René Dubos (1901-1982) acknowledged that the "concept of 'becoming' was obviously borrowed, perhaps unknown to Pasteur, from the *Werden* of the Hegelian doctrine." [2] It's noteworthy that one of the other speakers of this conference, Dr. Rasmus Haukedale, argued for moving beyond Kant's organismic perspective onto that of Hegel's. The concept of "dependent arising" is much closer to a dialectic approach of mutual becoming, in that it embraces fundamental interconnectedness across existential and conceptual domains of reality. That being said, dependent arising seems to imply that the origin of the causal relation of dependencies doesn't inherently possess this interdependent dynamic, thus it "arises." The core doctrine of Bhāgavat Vedānta, *achintya-bhedābheda*, along with the Hegelian dialectic, holds that the entanglement of dynamic dependencies is intrinsic to reality. This subtle and nuanced topic will not be elaborated here, but is relevant for further consideration.

References

1.   Kant, I, Bernard, J. H. (trans). 1914. *Kant's Critique of Judgement* § 64-65. MacMillan and Co. Retrieved from https://oll.libertyfund.org/titles/bernard-the-critique-of-judgement

2.   Dubos, R. J. 1950. *Louis Pasteur: Free Lance of Science*, pp 395-396. Little, Brown and Company.

"He [Louis Pasteur] reached the conclusion, as Claude Bernard had, that the mystery of life resides not in the manifestations of vital processes all of which pertain to ordinary physicochemical reactions but in the predetermined specific characters of the organisms which are transmitted through the ovum, through what he called the 'germ.'

[Quoting directly from English translations of Pasteur's original French writing] 'The mystery of life does not reside in its manifestations in adult beings, but rather and solely in the existence of the germ and of its becoming. . . .

'Life is the germ with its becoming, and the germ is life. . . .

'Once the germ exists, it needs only inanimate substances and proper conditions of temperature to obey the laws of its development ... it will then grow and manifest all the phenomena that we call 'vital,' but these are only physical and chemical phenomena; it is only the law of their succession which constitutes the unknown of life. . . .

'This is why the problem of spontaneous generation is all-absorbing, and all-important. It is the very problem of life and of its origin. To bring about spontaneous generation would be to create a germ. It would be creating life; it would be to solve the problem of its origin. It would mean to go from matter to life through conditions of environment and of matter.

'God as author of life would then no longer be needed. Matter would replace Him. God would need be invoked only as author of the motions of the world in the Universe.'

Like an obsession, there recurs time and time again through his writings, often in unpublished fragments, the statement that 'Life is the germ and its becoming.' The concept of 'becoming' was obviously borrowed, perhaps unknown to Pasteur, from the *Werden* of the Hegelian doctrine. This taught a logical or dialectic development of things according to which the whole world — spiritual phenomena, man, together with all natural objects — was the unfolding of an act of thought on the part of a creative mind."

3.   Han, S. M., Cottee, P. A., and Miller, M. A. 2010. "Sperm and Oocyte Communication Mechanisms Controlling C. elegans Fertility." *Developmental Dynamics* 239(5):1265–1281. https://doi.org/10.1002/dvdy.22202

"Sperm and oocyte fusion induces egg activation, a process that restarts the oocyte meiotic divisions, initiates embryonic development, and prevents polyspermy. Ovulated oocytes that are not fertilized in a timely fashion degenerate, essentially wasting the energy and resources necessary to produce them. [...] The molecular machinery that mediates sperm and oocyte communication is largely unexplored, with the exception of several marine species that are amenable to biochemical analysis."

4.   Zafar, M. I., Lu, S., and Li H. 2021. "Sperm-oocyte interplay: an overview of spermatozoon's role in oocyte activation and cur-

rent perspectives in diagnosis and fertility treatment." *Cell & Bioscience* 11(4). https://doi.org/10.1186/s13578-020-00520-1

5.   Hegel, G.W.F., and Baillie, J.B. (trans). 1807. *Phenomenology of Spirit* §276. GWFHegel.org. https://www.gwfhegel.org/Phen-Text/compare.html

6.   Hegel, G.W.F. and Miller, A.V. (trans). 1830. *Hegel's Philosophy of Nature* §370 Remark. Oxford University Press.

7.   Hegel, G.W.F. and Miller, A.V. (trans). 1830. *Hegel's Philosophy of Nature* §370 Zusatz. Oxford University Press.

8.   Doctor, T.; Witkowski, O.; Solomonova, E.; Duane, B.; Levin, M. 2022. "Biology, Buddhism, and AI: Care as the Driver of Intelligence." *Entropy* 24(710). https://doi.org/ 10.3390/ e24050710

9.   CSAS. 2020, September 30. "Turing Machines, Mutual Understanding and Dependent Arising." https://apparent-selves.org/blog/
turing-machines-mutual-understanding-and-dependent-arising/

2

Rasmus Haukedal

The Dialectics of Life: Towards a Unification of Biology and Philosophy

Rasmus Sandnes Haukedal is a postdoctoral researcher at East China Normal University. He earned his PhD from Durham University in 2023. In his thesis, he employed an organisational perspective to argue that the current expansion of evolutionary biology represents a dialectical turn. Building on this foundation, his postdoctoral project explores philosophy of mind, developing an organisational understanding of cognition. During his PhD, he was a research fellow in the Marie Skłodowska-Curie programme, 'Real Smart Cities', and he co-convened the reading group at the Centre for Culture and Ecology. He is currently an editorial member of Dialectical Systems and Salongen. He published Agency and Organisation: The Dialectics of Nature and Life (2025) with Springer Nature Link.

Dr. Rasmus Haukedal began his presentation with an overview, which included four sections: (1) Background: The Extension of the Modern Synthesis, (2) The Organisational Perspective, (3) The Return of Dialectics of Nature, (4) Conclusion: Another Kind of Naturalism,

Another Kind of Science? Before diving into these sections, he offered a description of modern science's present understanding of nature as a mechanical, dead, external, and passive reified background, drawing from historian of science Dr. Carolyn Merchant's *The Death of Nature* (1980). The morbidity of this description framed the need for the presentation, which argued for a new kind of naturalism that embraces a dialectic approach to thinking about natural phenomena.

PART 1 | Extending the Modern Synthesis

As commonly interpreted, the Modern Synthesis (MS) is the combination of Charles Darwin's theory of natural selection driving biological evolution and Gregor Mendel's theory of inheritance. Darwin tried to explain how biological changes happen, while Mendel explained how these changes are consistently passed on to new generations. Thus, evolution is sometimes defined as the change in allele frequencies or the frequency of particular gene expressions within a population. Historian of science Dr. Betty Smocovitis referred to the MS as "a delicate balance" because it was inspired by the positivistic ideal of unifying the natural sciences, but, due to positivism's particular biases, this project risked reducing biology to physics and chemistry. According to Dr. Haukedal, this delicate balance was achieved by claiming that biology was an emergent property of physics and chemistry. Thus, while the former relies on the latter to some extent, it is not completely reducible to them.

Being that this presentation was based on Dr. Haukedal's 2022 dissertation titled "Agency and Organisation: The Dialectics of Nature and Life" (https://etheses.dur.ac.uk/14893/), he shared his discovery of the inconsistencies within the deeper body of work concerning the foundation of the MS, due to varying disagreements among scientists, as well as a difference between the USA and UK version of the Modern Synthesis. Thus, Dr. Haukedal suggests thinking of the MS as a philosophy of nature rather than a scientifically validated research framework, given its lack of cohesion. Due to the inadequacies of this particular philosophy of nature, he offered a more coherent one.

Dr. Haukedal shared points of agreement with, and some of the many challenges to the MS. Pulling from Dr. Smocovitis's work, the three main points of agreement are:

1. The primacy of natural selection
2. The gradual rate of change, "operating at the level of small individual differences"
3. A "continuum between microevolution and macroevolution"

The challenges included epigenetics, niche construction, plasticity-driven evolution, symbiosis, facilitated variation, natural genetic engineering, extended inheritance, multilevel selection, and (ecological-)evolutionary-development biology. Before moving on to the next part of the presentation, he asked whether it is reasonable to think that extending the MS to address such challenges is even possible. Traditionalists say that the MS has already been expanded to accommodate past challenges and can continue to do so without needing a radical reformulation of the theory. Dr. Haukedal acknowledged the legitimacy of this claim, but pointed out that this approach is what led to the aforementioned inconsistencies being superficially unified, leading to deep incoherencies in the MS. Revisionists, on the other hand, who, to varying degrees, see the need for a significant change in 21st-century biology's evolutionary theory, feel that expansions of MS often lead to undermining the importance of the new phenomena that are being absorbed.

A helpful table comparing the Modern Synthesis (or Standard Evolutionary Theory – SET) with what has come to be called the Extended Evolutionary Synthesis – taken from Dr. Lynn Chiu – was provided on the last slide of the first part of this presentation:

SET assumptions	EES assumptions
The pre-eminence of natural selection	Reciprocal causation
Genetic inheritance	Inclusive inheritance
Random genetic variation	Nonrandom phenotypic variation
Gradualism	Variable rates of change
Gene-centered perspective	Organism-centered perspective
Macro-evolution is explained by microevolutionary processes	Macro-evolution is also explained by additional evolutionary processes, e.g., developmental bias and ecological inheritance

PART 2 | *The Organisational Perspective*

The common theme across the varying perspectives engaging with an extended evolutionary synthesis is that the organism is at the center of evolution, rather than molecules like DNA. This is where the organizational approach picks up. Dr. Haukedal describes it as the modern form of organicism and as a theoretical systems biology that seeks organizational principles governing the relationships among biological parts and between biological parts and the organism as a whole. This organizational perspective did not arise directly from new empirical research but rather aims to provide a more comprehensive framework for updated descriptions of phenomena already examined.

The first organizational concept discussed by Dr. Haukedal was constraints, defined as "boundary conditions which selectively harness matter to perform functions," often taking the form of physical factors acting as organizing principles. He distinguished between holonomic and non-holonomic constraints, where an example of the former might include phenomena such as gravity and electromagnetic fields, which appear independent from what they constrain, while phenomena like cell membranes exemplify the latter, which are generated from that which they constrain. Constraints are distinguished from processes, which are the content being constrained, like a river bank and the river itself. Constraints are not affected by the processes that they constrain. Since any determinate process implies constraints, and an organism is a totality of biological processes, constraints apply at every level of organization. Constraints do not only limit, but they also enable.

The organizational perspective holds that there is a categorical difference between non-living and living systems, and that the principles governing the former are insufficient to fully explain the latter. This perspective also offers a more dynamic view of biological function from the MS, where the MS relies on natural selection as the mechanism underpinning function, and the organizational view treats biological functions as contributing to the maintenance of organization. Each part of the biological system contributes to maintaining the system as a whole, which is called organizational closure.

Dr. Haukedal described three interconnected theoretical principles from the organizational perspective:

1. Default state: The default state of cells (living systems) is constitutive proliferation with variation and motility. This contrasts with the default state of non-living systems, which is rest or inertia. It is the primary generator of variation in living systems.
2. Variation: The MS views this as random, as it results from mutations rather than goal-directed biological processes. The organizational perspective views variation as arising from biological processes at all levels of organization. It also concerns "extended criticality," or changes from one qualitative state (a critical point) to another. "Extended" refers to the constant state of dynamic change that organisms undergo to maintain themselves. Regular "criticality" can simply refer to predictable changes in lifeless physical objects that occur at specific junctures, followed by pe-

riods of relatively little change, as opposed to the largely unpredictable, ongoing changes of living systems.

3. Organization –

   a. Organizational differentiation: parts of the organism differ from each other, and they all contribute to the functioning of the whole. Since the contributions of the parts are specialized, the parts of an organic system are not interchangeable, as with the parts of a mechanical system. Differentiation distinguishes organisms from dissipative systems like flames or tornadoes because all the parts of such systems serve a single function.

   b. Organizational closure: The closure of constraints implies that "the existence of each constraint depends on the existence of the others, as well as on the action that they exert on the dynamics. In this kind of situation, the set of constraints realizes self-determination as organizational closure."

      i. Organizational closure is coupled with thermodynamic openness, i.e., an exchange of matter and energy with what is other than the system. Dr. Haukedal points out that without this coupling, there would be no way to perpetuate the organization. Thus, an organism's closure does not imply independence from the environment; rather, it underpins the potential for relationality between them. It is not closure in an absolute sense, because this (a lack of exchange between organism and environment) would imply death. This offers a new kind of naturalistic framework for understanding self-determination through self-constraint or self-maintenance, showcasing the reality of teleology through the organism's goal-directed organizational activities that perpetuate itself by integrating the environment. Some have referred to this view of the organism as a Kantian whole, drawing from Immanuel Kant's *Critique of Pure Judgment*, where the organism is the cause and effect of itself.

Dr. Haukedal argued that the organizational perspective is beyond Dynamical Systems Theory (DST) – which he also defended as still relevant – because it acknowledges that the possibility space of biological activity cannot be effectively calculated due to the novelty of biology's tendency to find unanticipated solutions to unexpected problems. DST requires that you predefine all potential activities to calculate their probabilities, but the organizational perspective recognizes that these are not given in advance.

PART 3 | *The Return of Dialectics of Nature*

The third section of the presentation aimed to explain how the dialectic philosophy of G.W.F. Hegel (1770-1831) can inform the organizational perspective, and in turn, how the organizational perspective might clarify some aspects of Hegel's work. Significantly, Dr. Haukedal began by acknowledging that no other modern thinker has as many divergent interpretations as Hegel. Dr. Haukedal explained that the view he espouses is becoming more prevalent; however, it is "not the school book version of Hegel." Hegel's philosophy is experiencing a renaissance of resurging interest, including his philosophy of nature. The view of Hegel's philosophy of nature that Dr. Haukedal will draw on overcomes persistent misunderstandings of Hegel's overall conception.

Dr. Haukedal introduced two opposing interpretations of Hegel's philosophy of nature:

1. Hegel's conception is purely *a priori*, such that he offered armchair speculation about reality and then tried to use empirical scientific findings to justify these ideas.

2. Hegel's conception is empirical in that he used sensuous observations of reality to develop his philosophy, but present empirical findings have advanced beyond those in Hegel's day; thus, his philosophy of nature can be discounted as outdated.

Both of these interpretations overlook Hegel's dialectically holistic worldview, which sees science and philosophy as distinct yet continuous. The explanations offered throughout Hegel's philosophy of nature emerge from the interplay of science and philosophy. Science provides the content with which philosophy must work. Hegel says that "the empirical sciences have redded this material for philosophy," and we know that he worked alongside scientists while writing his philosophy of nature. Dr. Haukedal asserted that Hegel's work prefigured modern research.

Drawing from Andrea Gambarotto and Luca Illetterati's work in *Hegel Bulletin* Volume 41 - Special Issue 3 - December 2020, "Hegel and the Philosophy of Biology," Dr. Haukedal described three points of convergence between Hegel and the organizational perspective: (1) the nature of biological organization, (2) the role of teleology in biological explanation, and (3) the relation between life and cognition. He also brought out two other points: (4) reciprocal interdependence between organism and environment, and (5) Hegel's notion of randomness or contingency. Due to time constraints in the presentation, Dr. Haukedal explained that he would not be able to substantiate all five claims, but they are worth mentioning nonetheless.

Introducing the concept of dialectics, Dr. Haukedal mentioned three *informal* laws of dialectics articulated by Friedrich Engels: (1) the law of the transformation of quantity into quality (and vice versa), (2) the law of the interpenetration of opposites, and (3) the law of the negation of the negation. These should be viewed more as tendencies than laws, as the subject of dialectics cannot be reified into such static principles as laws of nature. These merely help introduce dialectics. Dr. Haukedal likened the first law to the phase change from water to ice, the second to the relation between organism and environment, and the third to a principle of change that explains how new properties are produced through the interchange of opposites. He further explained that these laws are interconnected and that, especially the first and third, are concerned with emergence. Before proceeding, Dr. Haukedal acknowledged that the term "emergence" is often used ambiguously and clarified that it refers to the irreducibility of higher levels of organization that depend on the activity of lower levels, where the higher levels are governed by laws/principles that cannot be reduced to those of the lower levels despite the causal connection. Emergence concerns both upward and downward causation.

The historical connection between dialectics and science was briefly explored, describing how dialectics inspired the original organicists of The Theoretical Biology Club, how ecology was formulated through its application, and how contemporary biologists such as Stephen Jay Gould, Richard Lewontin, and Richard Levins employed dialectical principles. From this perspective, organisms are not just objects, but are also subjects.

In Hegel's view, the living organism is "essentially process," one of self-differentiation and simultaneous unity-in-multiplicity. He articulated three stages of organism:

1. Geological organism: This refers to the planet as a precondition for life, not to the planet as such being alive.
2. Plant organism: Plants lack the differentiation of parts. Plant parts lack a specific identity, unlike specialized animal organs.

Plant parts regenerate, and a whole plant can itself generate from the part of another.

3.   Animal organism: The parts of an animal become deeply integrated members or participants that have particular functional roles in relation to the purposes of the whole, where the whole and its participants reciprocally rely on each other. These functions are very specialized and not so easily regenerated, if at all, as with plants.

    a.   *The last two bullets have been elaborated in Dr. B Madhava Puri's presentation, which is also included in these conference proceedings*

Dr. Haukedal draws a connection between Hegel's animal organism and the previously described organizational perspective's notion of differentiation.

Moving on to niche construction, Dr. Haukedal reiterated the description of organisms' self-maintenance through the mediation of their environment and characterized it as a process of attunement. Relevantly, Hegel holds that we cannot discuss an environment without including what occupies it (organisms) in our discussion. This principle is also found in Lewontin's work. So, the organism simultaneously attunes itself through and to the environment. The environment offers a possibility space with a set of potential resources, and each unique organism utilizes a subset of those possibilities as is relevant to their particular nature. This actualized subset of resources used by an organism – those things explicitly affected by it and those that affect it – is its niche. In other words, a niche encompasses those things in the world that correspond to the nature of the organism itself; only these count as relevant to its decision-making and therefore guide its action. The possibility space of the environment is determined into a particular world, a niche, through habit formation. This is a negation of the immediacy of the environment, and, mediated through organismic habituation, the environment is assimilated into the being-for-self of the organism, just as food, air, water, and sunlight are metabolically transformed from their immediate forms into whatever is necessary for the organism's self-maintenance. Dr. Haukedal offered a valuable passage from Hegel's *Science of Logic*, which beautifully explains the dialectics of niche construction:

> "'In so far as the object confronts the living being at first as something external and indifferent, it can affect it mechanically, but without in this way affecting it as a living thing; and in so far as it does relate to it as a living thing, it does not affect it as a cause but it rather excites it. Because the living being is an impulse, externality impinges upon it and penetrates it only to the extent that in principle it is already in it; hence the effect on the subject consists only in that the latter finds that the externality at its disposal accords with it.' (Hegel 2010: 685)"

He moved on to discuss three senses of organisms' subjectivity. The first sense, as previously elaborated, regards the organism as the subject or concrete unity of varying parts. The second sense regards autonomy, or the ability to interrupt mechanical causes in the process of attuning or assimilating the environment to itself, where, despite being constrained by natural forces, it assimilates or sublates them into its functioning. The inner purposiveness driving this self-determination is intimately tied to self-preservation. The third sense involves lack and self-feeling. Dr. Haukedal clarified that we're talking about self-feeling as common throughout all living entities, not a highly developed consciousness of self-feeling. Hegel explains that a sense of lack is fundamental to all organisms, and that the experience or feeling of this lack, along with the resultant drive to overcome it, motivates them to act in their environment.

One of the takeaways of Dr. Haukedal's presentation is his suggestion that we move beyond Kant's holistic thinking to Hegel's dialectical thought. Kant posited that teleology (purposiveness) is merely a heuristic, a useful tool for understanding nature that can ultimately be discarded, which Dr. Haukedal found similar to modern biologists' reluctance to explicitly acknowledge purposiveness despite the abundance of goal-directed biological activity. Kant held that teleology lacked ontological reality. Hegel said that Kant did a "great service to philosophy" by acknowledging teleology to some extent, but Hegel also noted tension in Kant's position due to its lack of a way to unite ontology and epistemology. Our knowledge of the world as purposive (epistemology) can not be divorced from the actual being of the world (ontology). Hegel overcame this tension by explaining that living beings are part of the world, sharing ontological similarities with it, which is why we can know it. Our concepts share structural similarities with the world; teleology or purposiveness helps us understand the world because the world actually embodies that quality. Nature is intelligible because it intrinsically possesses rational organizational principles that are also possessed by thought. Thought recognizes these aspects of itself in nature.

As Dr. Haukedal mentioned, this train of thought is related to Hegel's distinct discussion of the Concept and its embodiment in the organism. The Concept is the self-organizing rational impetus that drives the expression of what is internal to manifest externally, and integrates what is external with the internal. It is embodied in the organism's activities and differentiated through its judgments. This is not merely an outward movement of imposing concepts on the world. The concepts themselves emerge from the fundamental ontological ground shared by the subject and object. So, there is an external element in Hegel's discussion of the Concept, such that the Idea is when the ideal (logical determinations) and the real (natural existence) meet, or the unity of the concept and its reality. Moving our philosophy of nature from Kantian wholes to dialectical wholes entails adopting Hegel's view of teleology and its implications for the interrelatedness of ontology and epistemology.

Wrapping up part 3 of his presentation, Dr. Haukedal brought our attention to the impotence of nature. Hegel says that "nature is the Idea in its otherness," or "the Idea shown in a fussy way." This important insight of Hegel shows why his conception is not about deducing nature from logic; rather, he argues that our concepts will never *perfectly* map onto the world. Logical categories are presented in a fragmented manner in nature, due to individuals in nature being contingent or relative to what is other than them. Thus, an individual, as such, could never completely embody our concepts of it, which spill over into unobserved potentiality and relationality of the individual. This does not contradict the earlier argument of Hegel's unity of ontology and epistemology. Rather, it means that our extensive knowledge of nature is finite, and that phenomena can appear in unexpected and inexhaustible manifestations.

PART 4 | *Conclusion: Another Kind of Naturalism, Another Kind of Science?*

Hegel's naturalism breaks down the barrier between subjects and objects, between mind and matter, thereby eroding the boundary between culture and nature. Some have characterized this as a "non-naturalistic [in the traditional sense] kind of naturalism." Significantly, Hegel said that the contingency of nature places strict limits on philosophy *and* on science. Thus, Dr. Haukedal advocated that we should not enter into science with the presupposition that nature can be completely known, as this fails to recognize nature's ongoing, living, and dynamic quality. He inquired whether the practice of science could remain as it is within this new kind of naturalism.

Dr. Haukedal concluded with two questions. The first aimed at inspiring self-perpetuating continuity between theory and practice, which equally applies to philosophy of nature, philosophy of science, and the science of philosophy: "What kind of science do we want, and how can we get a kind of critical science where we apply the principles that result from scientific research to science itself?" The second inquiry concerned the future role of philosophy in science, which necessarily involves deep engagement with the history of both science and philosophy to inspire discoveries and facilitate cross-disciplinary innovation. Hegel's insightful approach to science and philosophy is instructive for advancing in this way.

3

B. Madhava Puri

A Conceptual Foundation for Deeper Insights into Living & Nonliving Systems

Bhakti Madhava Puri received a PhD in Quantum Chemistry from Georgetown University and worked as a postdoc at the National Bureau of Standards in Washington DC. He published technical papers in The Journal of Chemical Physics. After encountering the measurement problem, he turned to the Indian school of Bhakti-yoga to learn about consciousness. Dr. Puri started GWFHegel.org, was the Serving Director of the Princeton Bhakti Vedanta Institute, as well as the visionary behind this annual conference series since 2013.

PART 1 | *The Significance of Mechanistic & Organicist Conceptual Frameworks*

Dr. B. Madhava Puri's presentation began by distinguishing between a mechanical account of reality, which posits matter as fundamental, and an organic account of reality that corresponds to spiritual insights in the philosophies of G.W.F. Hegel and ancient India's

Vedanta. Both the mechanistic and organic perspectives regard reality as a system, but they differ in their views on the nature of the necessity governing the system's unity, and on the relationships between the parts of the system and the whole, as well as between the parts themselves. These differences are grounded in particular epistemological and ontological views, such as reductionism or holism, and positivism or embracing negativity. Because such views shape one's worldview, they also influence opinions on how science should be practiced and which sciences are regarded as fundamental. This further affects how research agendas are formulated and how conclusions are drawn regarding essential topics such as the purposiveness at various levels of organization in nature, distinguishing between life and non-life, and the origin of life.

To frame the forthcoming ontological account of the nuanced continuum of bodies encountered in nature, Dr. Puri began by describing two poles of this continuum, characterized by general qualities that distinguish mechanical and organic systems.

Traditionally, mechanical systems are understood through the epistemologies of reductionism and positivism, where the former holds that a whole is nothing more than the sum of its parts, and the latter holds that things are real only if they physically exist and are empirically observable. In the classical Newtonian sense, the unity of a mechanical system is contingent upon external agency. In other words, forces outside of the system act to aggregate its components. Because mechanical systems are governed by external necessity, their components are indifferent to or independent from one another, even though their collective work powers the system. Lacking an integral relationship with one another and with the whole, mechanical components are easily replaceable, as exemplified by human-made artefacts such as clocks or other machines. Altogether, this is the conception underlying modern science's notion of matter. Dr. Puri further described matter as an *"existent static being,"* highlighting science's view of the material world as a collection of things with ossified, fixed identities. Alternatively, as will be seen, the organic perspective views the world as a *processual, dynamic becoming*.

The Western world has come to label the study of the laws and principles that describe the motion and interactions of the parts of mechanical systems as the science of physics. While mechanical systems clearly exist in nature – as seen in rock formations, flowing water, and planetary orbits – it would be a categorical error to reduce the entirety of nature to them. Nonetheless, modern science presupposes that physics is the most fundamental science, and by imposing the knowledge gleaned from physics as the primary framework for studying biological systems, we end up reducing the science of biology to a mere redundancy of physics. This is a great disservice to the kingdom of life, however. Historically, the consequences of reducing biology to physics have included the conclusions that living entities are inherently purposeless, that they are nothing more than aggregates of nonliving material building blocks, and that the origin of life is matter (the theory of abiogenesis). Due to results in the contemporary research of 21st-century biology as seen throughout these conference proceedings, which recognize the profoundly and irreducibly integrated nature of organisms and the central role of purpose permeating activity and functionality at all levels of organization within the organism, there has been a resurgence of interest in fresh iterations of the organismic or organicist thinking that preceded modern mechanistic science.

Dr. Puri explained that instead of an ossified world with isolated parts externally unified, the organicist view sees the world as a dynamic process of becoming whose constituents are fundamentally interdependent and internally related. This perspective regards biology as the primary natural science and holds that physics and chemistry should be studied in relation to it. Becoming involves several moments that necessarily contextualize existence: the coming-to-be and ceasing-to-be of a thing. Whatever exists is coming from something (the formal aspect of cause) and is going toward something (the final aspect of cause, being-for, purposefulness, or teleology). These diachronic relationships coexist with synchronic relations, such as the material and efficient causal relations characterizing mechanical systems, where diachronic refers to the totality of a process, and synchronic refers to a single aspect or moment of the process abstracted from the totality. Since the mechanistic view is positivistic, it accounts only for positive, existing objects, static and abstracted from the context of their development. But the organicist approach is comprehensive enough to encompass being as one of the moments in the development of becoming, along with being's opposite or negation, nonbeing. This dynamic process necessitates consideration of the negative as well as the positive, where development is a process of negating what a thing is so that it can become what it is not. This coming-to-be and ceasing-to-be is dialectical in the Hegelian sense because the negation of one stage of development emerges from the stage that is being negated. It is a self-propelling negation that drives the continuous actualization of the organic system. Dr. Puri emphasized that without considering the negative, all assessments of developmental processes must remain superficial.

Hegel's systematic philosophy offers a framework for conceptualizing nature and the world as *processual dynamic becoming* rather than *existent static being*. The latter is contextualized within the former as a snapshot of its ongoing activity. Hegel uses the term *Aufheben*, often translated as "sublate," to describe the simultaneous negation of the explicit form (where negation does not entail annihilation) and the preservation of the implicit function of underdeveloped concepts within more comprehensive and accommodating ones. This will be seen in his conception of mechanical, chemical, biological, and human bodies, which follows. The term "Organic Whole" was used by Hegel and authorities of Bhagavat Vedanta philosophy, such as Srila Bhakti Rakshak Sridhara Dev-Goswami Maharaja, to describe the Absolute Truth, which is understood as the most comprehensive concept or Absolute Idea that sublates and contextualizes all other concepts. When the Absolute Idea is known to possess self-consciousness, as Hegel explained, its similarity to the Vedantic conception of Brahman is revealed.

Following the example given by the ancient *Vedanta-sutra*, which begins with the aphorism "*athato brahma jijnasa* – Now, therefore, inquire about Spirit (Brahman)," Dr. Puri advocated reimaging modern science in a manner that starts with the Organic Whole and then seeks to understand how matter is one integrated aspect of the whole, along with other aspects such as life and consciousness. It is necessary to clarify that calling reality the Organic Whole is not claiming that it is a biological organism. Just as matter is an immature moment of the Organic Whole's development, organism is a slightly more mature stage. Just as matter is not the foundation of reality, neither is organism. The word organism is derived from the Greek *organon*, meaning "instrument," which rightly points to a greater whole or totality beyond the individual organism, to which the individual is instrumental, i.e., functionally related.

The next part of Dr. Puri's talk presented a conceptual framework that delineates the logical distinctions and relations among mechanical, chemical, biological, and human bodies, grounded in Hegel's philosophy. This framework helps to distinguish between life and nonlife not as mutually exclusive binary categories, but as necessary logical distinctions of the spectrum of embodied consciousness.

PART 2 | *The Logical Development of Self-Other Relationality among Natural Bodies*

The logical distinctions between mechanical, chemical, biological, and human bodies are determined by the interdependent dialectic dynamic between *self* and *other* or, more concretely, an individual and its environment. The identity of each kind of natural body is found in its other and the extent to which a particular body unites with otherness, where this union is the claiming of its identity, determines its kind. The limitations that one type of body encounters while attempting to connect with its other sets the necessary standard for the determination of a new type of body. This conceptual development logically connects mechanical, chemical, biological, and human bodies, where humans are the most capable of integrating otherness and thus truly claiming their identity.

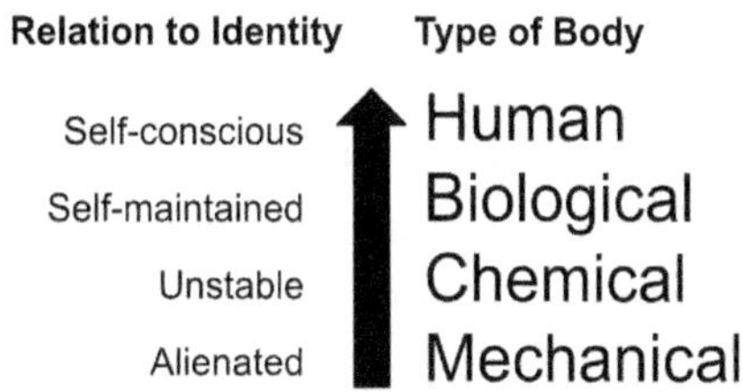

The logic of this approach is empirically justified, as the principle that natural bodies inherently require something external to themselves for their existence is evident in phenomena such as amino acids. Amino acids are fundamental constituents of proteins throughout living bodies that support biological processes such as growth, healing, and digestion. Out of the 20 amino acids required for protein synthesis, 11 "nonessential amino acids" are made by the living body, and the remaining nine "essential amino acids" cannot be made by the body and must come completely from diet. This is one empirical example of how the self constitutionally relies on its other.

This logical development of self-other relationality across kinds of natural bodies begins with the concept of externality, as discussed in Hegel's *Encyclopedia of Nature*. This complements the Vedantic conception of the *bahiranga-shakti* (external potency) or *maya-shakti* (illusory energy) of the Absolute. In an immediate sense, individual bodies appear to exist externally to one another, although a deeper analysis of nature reveals that everything is intimately and intrinsically connected. In the case of natural bodies, externality refers to the identity of the body being outside of itself, i.e. bodies are necessarily defined by their relation to what is other than themself. The degree to which a body successfully overcomes externality and claims its identity determines its superiority to the other kinds of natural bodies, where the standard for superiority is the prolonged capacity for assimilating otherness. This development reveals a spectrum of self-other integration, in which inferior bodies cannot sustain it, thereby giving rise to the logical necessity for a body that can. The human body is superior because it can permanently maintain the unity between self and other, grounded in the exclusively human capacity for self-reflective thought (distinct from lower-order cognition in cells, plants, and animals). Ultimately, this capacity for thought manifests as the prospect of attaining self-knowledge and love, in which one sees oneself in the other. Dr. Puri then systematically explained the nuances of how the human body necessarily emerges to overcome the inadequacies of mechanical, chemical, and nonhuman biological bodies, thereby sufficiently integrating otherness and claiming its identity. His development is summarized in the chart below.

	Necessity of Existence	Inadequacy	Relation to Identity (identity of self lies in the other)
Human	Through the capacity for thought, establishes unity that does not require the annihilation of difference	Unity of self and other is *not disturbed*, due to the accommodation of unity-in-diversity	Self-conscious
Biological	Establishes continuous unity where otherness is consumed and assimilated with the unchanged self	Unable to assimilate mate during reproduction, disturbing unity	Self-maintained
Chemical	Establishes an initial unity with otherness as a compound	Polar nature causes perpetual bonding, disturbing the unity	Unstable
Mechanical	The most basic existence of a spatially extended body where self and other remain external to each other	Dependent on otherness yet repels the other when encountered, preventing unity	Alienated

Mechanical Bodies

According to René Descartes, a corporeal body is an extended sub-stance (*res extensus*), or more concretely, spatial extension. In a spatially extended body, each point is external to the others; that is, each point occupies a distinct location with different coordinates. Thus, externality is at the root of extension. These coordinates are all contained within a single body, as defined by its surface. The surface is the boundary that demarcates the self from the other, or the individual from the environment. Because the surface's unique figure or shape is the most fundamental quality of a mechanical body, the shape constitutes its identity. As will be seen, the shape of the body is determined by the environment; hence, we can say that the identity of mechanical bodies (and all kinds of natural bodies) lies in their other.

Before proceeding, Dr. Puri further clarified the definition of mechanical bodies. Bodies determined purely by spatiality are mechanical. Interactions between mechanical bodies are external relations in which individuals do not change, and where there is no internal impetus driving their relationship. Thus, mechanical unities are just aggregates formed through external necessity, as explained in the first part of the presentation. This is how modern science typically views mechanical systems. A mechanical body remains the same whether or not it interacts with other mechanical bodies, which indicates an indifference to otherness. This immediate indifference creates tension with the mediated fact that the identity of the mechanical body depends on its other. This discrepancy renders the conceptual framework of mechanism inadequate for complete self-expression, giving scope for the framework of chemism (relating to chemical bodies).

We observe that the orientation of a gear's teeth is determined by its relation to another gear so that the teeth interlock and rotate, and the shape of a pool ball is determined by its relation to the pool table that it rolls across and other balls that it collides with and ricochets against. The figure of a rock is a byproduct of erosion, indicating a mechanical relationship with the movement of air and water. In each case, the mechanical body is directly determined by elements in the environment; however, upon contact with those elements, no meaningful transformation or exchange occurs. They just collide and rub against eachother's surface. The mechanical body cannot claim its identity through the other; there is no integration between self and other, which shows a very limited capacity for self-expression. Coupled with the aforementioned discrepancy between immediate indifference to and mediated dependence on otherness, the mechanistic framework proves inadequate to describe the totality of observed natural phenomena. Rather, its explanatory value depends on its contextualization within a more coherent conceptual framework, within which it is a valid yet incomplete perspective. Thus, certain aspects of chemical, biological, and human bodies can be described mechanistically, but this approach will not fully explain these more sophisticated phenomena.

Chemical Bodies

In addition to having all the properties of mechanical bodies, chemical bodies overcome the inability of mechanical bodies to integrate with otherness, such that they remain indifferent to each other and do not transform through relationships. This is because chemical bodies intrinsically possess polarity, a natural affinity for attracting and bonding with one another, and thereby producing compounds that are distinct from the reactants. This transformation of chemical reactants into products demonstrates a degree of integration between self and other that is not found among merely mechanical bodies. The bond is only stable up to a certain extent, however. It can (1) be broken, and each reactant can return to its previous state, or (2) the tendency to attract and react compels the newly formed chemical compound to perpetually react and transform with other reactants in the environment. In either case, the unity between self and other is disturbed, indicating that genuine assimilation of otherness never occurs. As will be seen, this is what distinguishes chemical bodies from biological bodies. Biological bodies assimilate otherness into the self through consumption and metabolic activity.

Dr. Puri offered empirical examples of chemical bodies behaving as just described, before moving on to biological bodies. To demonstrate how chemical compounds continue to perpetually bond, he considered a water molecule (H_2O). Although water is considered a very stable product, it may still react with other chemicals to become a hydrate. Here, the other chemical absorbs water onto its surface or into its structure. A familiar mineral hydrate is Epsom salts (magnesium sulfate heptahydrate – $MgSO_4 \cdot 7H_2O$), used in therapeutic baths. Epsom salts are chemically bonded, thus a new chemical body has formed. As the next logical stage of development for natural bodies following mechanism, chemical bodies also exhibit lower-level mechanical relations among themselves, such that two chemical bodies can combine without becoming chemically bonded. Methane hydrates ($CH_4 \cdot 5.75H_2O$) are an example, where a methane molecule (CH_4) becomes encaged within interlocking water molecules. This produces an ice-like substance that forms under low-temperature, high-pressure conditions, such as in the deep ocean. Ultimately, the inadequacy of a chemical body's self-other relationality stems from its constant flux, given its fundamentally polar nature. Chemicals overcome the limitations of mechanical bodies, but their unity does not endure and is less stable than that of biological and human bodies.

Biological Bodies

Biological bodies are distinguished through their self-determined ability to assimilate otherness with themself. Recall that chemical bodies form compounds that can either be broken or perpetually bond, which is why they are deemed less sophisticated than biological bodies, where the standard for sophistication is the capacity for an enduring unity between self and other. Biological bodies, on the other hand, subsist through the absorption of otherness. When grass consumes sugar, or a cow consumes grass, only the food is transformed, not the consumer. The food's original constitution disintegrates into basic units that are integrated and repurposed by the consumer's body to fuel biological processes. Whether eating food, breathing air, or absorbing sunlight, biological bodies maintain their integrity by assimilating otherness.

After describing how the integrity of the self-other relationality in biological bodies surpasses that of chemical bodies, where the integration of self and other never persists in chemical relations, yet in biological relations, the consumer fully and irreversibly integrates food with itself, Dr. Puri provided an analysis of the varying degrees of integrity between plant and animal wholes. Immanuel Kant (1724-1804) offered a robust standard for determining the integrity of biological wholes. In his *Critique of Pure Judgement*, Kant described an organism as an entity where (1) the whole depends on the parts, (2) the parts depend on the whole, and (3) the parts depend on each other. This framework helps us discern the degree of integrity among various plants and animals. For most animals, the loss of a limb or an organ is a devastating and possibly lethal experience. Although there are exceptions where limbs and sometimes organs regenerate in fish and amphibians – even some reptiles can regrow their tails, and deer regrow antlers – limb and organ regeneration is not the norm in the animal kingdom. This indicates strong integrity among animal bodies, in which the whole and its parts are profoundly integrated and cannot exist without one another. Plants, on the other hand, do not demonstrate such strong integrity. Whether large or small, plant parts can be removed without significant interference with the functioning of the whole; the removed part may regenerate into a whole new plant, or it can be grafted onto a plant of a different species and survive. Plant cells are considered mainly "totipotent" because each cell contains information about the entire genome of the plant. Thus, under appropriate conditions, many plants can be grown from a single cell taken from an existing mature plant. This is called "somatic embryogenesis." The totipotency of plants underlies methods for propagating hundreds of plant species by taking leaf or stem cuttings from an existing mature organism.

The fact that plant parts are, in many ways, independent of the whole plant, and that the whole plant persists despite drastic changes to its parts, indicates a less intense integration between parts and whole. This lack of whole-part integration (as compared to animals) indicates less intense self-other relationality, because part of a plant's self can separate from it and develop into an other, i.e., a separate plant. Consider that removing the arm of a monkey or even of a regenerating starfish will not lead to a new organism growing from that arm. Further, we can not transplant a vital part of one animal onto or into another animal and expect a healthy organism. Current work on xenotransplantation tries to achieve this, but only as an experimental end-of-life treatment. Conversely, limbs of plants can be liberally attached to plants of a different species and continue to thrive. People have created so-called "fruit cocktail" trees through grafting. They can grow cherries, plums, and peaches all on one tree, even apples and pears, or lemons, limes, oranges, and grapefruits can be grafted onto a single tree. Grafting is possible for any plant with a cambium. These facts demonstrate that the parts of plants are less integral to the whole. A plant can be dismembered without missing a beat, but that calls into question whether such parts were members to begin with, or actually implicit wholes in and of themselves. In animals, the integration between the organism as a whole and its parts is tight, and the distinction between self and other is well defined. Thus, we observe potent self-expression in their struggle to survive and reproduce.

Lastly, to reinforce the distinction between plant and animal bodies, Dr. Puri highlighted differences in their capacities to assimilate the spatial aspect of the environment. These were more of an afterthought, intended to inspire further discussion elsewhere, and were not elaborated in the presentation. He explained that plants are stationary, remaining in a limbo between self and other (individual and environment), where spatiality is not assimilated. While they absorb otherness like sunlight, water, and minerals with themselves, plants do not assimilate the spatiality of the other into itself – it just remains a compound or conjunction of self and other – thus it fails to fully integrate otherness with itself. In contrast, animals are mobile, clearly assimilating spatiality and distinguishing themselves from the environment.

The conclusion is that although biological bodies in general demonstrate a superior capacity for self-other relationality, and therefore a more intense ability for self-expression than mechanical and chemical bodies, among biological bodies, animals demonstrate greater ability for self-expression than plants. Returning to the determinations of biological bodies in general and how their deficiency leads to the necessity for the human body, we can understand that living entities are eager to possess and assimilate what is other from themself in order to maintain their basic individual needs. In many cases, they literally absorb other-ness and transform it into themself, as seen through the consumption of food, air, and sunlight, demonstrating the claiming and maintain-ing of their identity. There is one critical aspect of the biological experi-ence, however, that similarly triggers taking possession of otherness, yet it cannot be satisfied by simply absorbing the other. This is reproduc-tion. Two mates satisfy this urge through sexual unity with each other, where both individuals remain as they are. Neither is transformed into the other. This frustrates the living entity's tendency to absorb other-ness, thus, sexual acts, at least for animals, can be quite abrupt and ag-gressive. To clarify, this sexual need of the biological body cannot be satisfied by assimilating otherness, so it interrupts the integration of self and other. This deficiency is overcome in the human body.

Human Bodies

The human body encompasses all of the previous kinds of bodies and relationships – mechanical, chemical, and biological – yet surpasses their ability for self-expression through self-other relationality. This is due to the exclusively human capacity for thought, where thought is a higher-order faculty than the cognition found in single cells, plants, and animals.

Despite humans being misleadingly described as "rational animals," Dr. Puri explained that no animal can be rational, because animals do not think. Although they possess mental activity and act on it instinc-tually, they are not conscious of their mental processes and cannot in-tentionally direct the logical development of such mental content. The lack of an animal's capacity to think is precisely what prevents it from overcoming the disintegration of self and other during sexual reproduc-tion. Nonhuman living entities have the capacity for cognitive expe-rience where they make decisions, cooperate, and solve problems, but only humans demonstrate awareness of cognitive experience while it's happening. Thought is the capacity for an intentional development of self-conscious experience. This self-aware experience enables us to inquire into our identity and purpose, which has given rise to the sciences, the humanities, and civilized life. Ultimately, thought proves to be the most sophisticated means of self-expression through self-other relationality.

Animals have no means of assimilating their mates. Their most sophisticated means of integrating otherness is physical consumption, where this is untenable for reproduction due to consuming the mate preventing sexual intercourse. Humans overcome this dilemma by assimilating their mate through thought. Sexual reproduction becomes a thoughtful and loving act – as opposed to an aggressive animalistic one – where the partners form a unity through their shared self-reflective affection. Animals instinctively feel that otherness (the environment) is for them, and they try to take and forcefully assimilate it. When this is unsuccessful, aggression arises. From a conceptual perspective, animals can only integrate otherness by destroying or dissolving differentiation. The other ceases being what it was and conforms to the needs of the animal. While this also applies to humans with respect to biological processes such as eating, breathing, etc, the sophistication of thought allows humans to integrate otherness *without* dissolving differences. Thought accommodates recognition of unity-in-diversity and identity-in-difference. Human partners can assimilate one another without destroying each other's bodies. This also extends to the relationship between humans and the environment. Humans assimilate nature not only by consuming it but by knowing it. Humans can even assimilate themself through self-knowledge. Hegel's philosophy offers a robust development of this line of thought, which is deeply complementary to the Bhagavat Vedantic perspective.

Conclusion

The conceptual development offered by Dr. Puri established a nested continuum linking mechanical, chemical, biological, and human bodies by examining how their distinct self-other relationalities accommodate varying degrees of self-expression. Here, bodies with a higher degree of self-other integration possess the qualities of those with a lesser degree, while the latter do not possess the qualities of the former. Thus, this non-reductionist framework can serve as a foundation for deeper insights into the ontological similarities and differences between nonliving and living systems, in which the logical necessity of progressively sophisticated bodies constitutes an ontological limitation of the bodies with more restricted relationality and integration.

Mechanical bodies prove to rely on their other in that the boundary of their determinate figure – which is their essence – depends on its relationship to other bodies. When these bodies meet, however, they only repel each other, such that any meaningful unity is not achieved. Chemical bodies unite with one another through chemical bonding, which arises from their fundamental polarity. This exact nature is what forces them to perpetually seek another body to react with, such that the unity of self and other is never maintained. Biological bodies can sustain their union with otherness. Plants and animals absorb and assimilate otherness in such a manner that eradicates the other's previous existence to perpetuate the self-maintained existence of the organism. Further, the whole-part relationality of plants proves less integral than that of animals, indicating that self-other integration is more fully realized in animals. Reproduction is the point at which nonhuman animal bodies demonstrate their deficiency in maintaining self-other integration, since they cannot physically absorb their mate and still engage in intercourse with them. The human being overcomes this dilemma by assimilating their partner, the natural environment, and even themselves, through thought, which establishes unity while preserving diversity.

The utility of Dr. Puri's conceptual framework lies in its potential to inform our thinking about the nuanced relationality between different kinds of individuals and the environment. The framework suggests a fundamental interdependent relation between an individual and the environment, and offers a foundation for considering how the degree of an individual's assimilation of the environment and consistent integration of its parts indicates its capacity to serve as a vehicle for the full expression of natural purposiveness, from rudimentary self-alienation in mechanical bodies to sophisticated self-expression through self-conscious thought in the human lifeform.

4

Bernd Rosslenbroich

Properties of Life: Proposal for an Integrative Concept

Bernd Rosslenbroich is head of the Institute of Evolutionary Biology and Morphology at Witten/Herdecke University, Germany. Research interests are patterns and processes in macroevolution, organismic and systems biology, and philosophy of biology and medicine. Author of On the Origin of Autonomy: A New Look at the Major Transitions in Evolution (Springer 2014), and Properties of Life: Toward a Theory of Organismic Biology (MIT Press 2023).

Dr. Rosslenbroich began by situating organismic biology historically as an approach that sought to understand life in a manner that avoided being reduced to either of the extremes of mechanistic thinking or vitalism. One early attempt to establish an integrative science of organismic biology was the Theoretical Biology Club at Cambridge University in the 1920s. Due to the emergence of innovative non-reductive scientific conclusions in 21st-century biology, new attempts have been made to update and establish the importance of organismic thinking. Dr. Rosslenbroich identified the work of three primary scientists who called for elucidation of organismic principles at the turn of the 20th century: American microbiologist Carl R Woese's "A New Biology for a New Century" (2004), British geneticist Paul Nurse's "Life, logic, and information" (2008), and British physiologist Denis Noble's *The Music of Life* (2006). With this background in place, Dr. Rosslenbroich explained that his work aims to provide a cohesive concept that can effectively orient organismic thinking, which, he believes, is currently lacking.

To establish a unified overview of the organismic perspective of life that captures the attention of the scientific community, Dr. Rosslenbroich offers a synthesis of living processes that have been observed through empirical research. He specifies that this is not an attempt to define life, which he feels has been largely unsuccessful despite considerable effort in this direction, and neither is he trying to describe the difference between nonliving and living. Synthesizing observations of organismic principles also leads one beyond physics and chemistry, establishing biology as its own ontological category. Dr. Rosslenbroich concludes this introduction by explaining the intentions behind his book, *Properties of Life: Towards a Theory of Organismic Biology* (2023), which serves as the backbone for his presentation. These intentions include:

- Proposing a comprehensive concept for organismic biology
- Arguing for an ontologically irreducible concept of the living
- Developing a background for dealing with nature in a more appropriate manner than mechanistic thinking

Dr. Rosslenbroich clarifies that his presentation aims to introduce the properties of life in a simple manner, which are more thoroughly elaborated in his aforementioned book, available open-access and freely at The MIT Press website (https://doi.org/10.7551/mitpress/14739.001.0001). These properties should be taken as starting points for research:

1. Interdependencies
2. Integrative systems
3. Autonomy
4. Agency
5. Processing of molecules
6. Processing of information
7. Processing of energy
8. Processes of shape
9. Time autonomy
10. Sensitivity and affectability
11. Subjective experience
12. Ability to evolve
13. Growth and development
14. Environmental interrelationships
15. Reproduction and death

1. Interdependencies || The fundamental method in science is to organize phenomena into linear chains of cause and effect. Naturally, the early pioneers of biochemistry at the beginning of the 20th century adopted this method. For instance, in the 1950s, the energy production of a cell was said to be linear from Succinat ◇ Fumarat ◇ Malat ◇ Oxalacetat ◇ Pyruvat ◇ Citrat. Later on, scientists understood that the process was actually a circular cycle, i.e., the citric acid cycle. Today, biochemical processes are taught as interdependent cycles with no initial cause or final effect, where each cycle is dependent on another, or a group of cycles is integrated into a network. Each cycle is a necessary condition for the other cycles to occur. In this regard, even the central dogma of molecular biology, which posits a linear causal chain from DNA to protein production, has been significantly challenged, as DNA cannot simply be seen as the first cause due to the essential feedback processes involved. In his book, Dr. Rosslenbroich termed these interdependencies as the principle of concurrency. This principle is essential not only for understanding organisms, but also for ecology.

2. Integrative systems ‖ Historically, there have been two very different approaches to understanding biological systems. Pragmatic systems biology attains vast data sets through reductionist approaches, then reconstructs the organism based on this information. Integrative systems biology, which traces its roots to the 1960s, considers biological systems as a distinct ontological category and defines such systems as relatively independent and stable entities that exert regulatory constraints on their subcomponents, which ensures the proper function of the system as a whole. Paul A. Weiss was a proponent of this latter approach. Pragmatic systems biology can be described as a bottom-up approach, whereas integrative systems biology emphasizes a top-down perspective. Dr. Rosslenbroich clarifies the interdependence of top-down and bottom-up biological activity, where the organism as a whole constrains the function and development of its parts, and the parts also actively construct the total organism.

3. Autonomy ‖ This property highlights a difference between an organism and the environment. Metabolism generates an imbalance within the organism with respect to the inorganic environment, where an energetic, genetic, and all-around organizational distinction is maintained. New possibilities arise for the organism because of this self-maintained distinction from the environment. In 2014, Dr. Rosslenbroich wrote an entire book titled *On the Origin of Autonomy: A New Look at the Major Transitions in Evolution,* dedicated to detailing the progressive degrees of autonomy from the environment that manifest as a result of evolution. One example is the transition from unicellular to multicellular organisms, where multicellular organisms have many more internal functions to regulate. Progressively complex organisms develop the capacity to stabilize and self-maintain their internal activity as distinct from the environment.

4. Agency || Agency is self-activity. The overall autonomous activity of the organism to maintain its life functions, which serve as the basis for the development of further activities. For example, metabolism, protein synthesis, and DNA transcription are all activities generated by the agency of the living cell. Life itself is agency or self-activity. In humans and animals, these activities take the form of organ systems and subsystems, such as the cardiovascular and circulatory systems. Despite the seemingly self-evident nature of these statements, they challenge conventional biological views that restrict activity to physicochemical properties. In the conventional view, biological functions are causally reduced to molecular energetic reactions and metabolic machinery. When viewed in light of the existence of agency, this conventional view is counterintuitive because agency only exists in an intact whole organism. Dr. Rosslenbroich highlighted that agency proper is not observed in inorganic nature by posing the question: "Is a stone falling down a waterfall really analogous to the salmon jumping up the waterfall?" Keeping this in mind, he specified that while metabolic processes and genetic information fuel and support organismic and cellular agency, they do not possess agency in themselves, in that they cannot actively control or determine anything within the organism. A living entity is more of a *process* than a *thing*, in that they are characterized by active self-regulation. This topic was the subject of Dr. Rosslenbroich's recent open-access article "Agency as an Inherent Property of Living Organisms" (https://doi.org/10.1007/s13752-024-00471-7).

5-7. Processing of molecules, information, and energy || Dr. Rosslenbroich discusses these three properties together due to their irreducible functional relation and dependency or concurrency on the agency of the organism as a whole. The integration of the properties of very specific molecules is very important to organisms. This is the topic of biochemistry. There is a reciprocal dependency where the type, timing, and location of molecules are actively determined by cells in response to an organism's needs, while the unique characteristics of the molecules are essential for cellular and organismic function. Similar to the processing of molecules, there is an interdependence between top-down (organism ◈ cells ◈ molecules) and bottom-up (molecules ◈ cells ◈ organism) activity regarding processing biological information. For instance, the cell determines which section of a DNA molecule is transcribed, i.e., which pieces of information encoded in the genome will be utilized, as well as when this is executed. Organismic and cellular information processing are cognitive activities; some degree of intelligence is present. Regarding energy, cells organize the processing of energy in small units of ATP (transporting energy), which underlie and support all other cellular functions. Agency is generated by molecules, information, and energy, and it cannot be reduced to just one of them. It integrates and regulates these interdependent properties. At the same time, the system itself is not mysteriously beyond these components; rather, it relies on them for its maintenance.

At timestamp 26:21 in the recording of this presentation, Dr. Rosslenbroich acknowledged scientists' effort to study molecules, information, and energy as individual entities, but then emphasized the necessity of reintegrating these properties – recognizing their concurrency – into the context of the organism as a whole. He says that only after this reintegration can such studies come closer to the truth of an organism. This comment feeds nicely into a discussion that happened at the end of the conference. During the panel discussion, from timestamp 1:16:27-1:19:15 in the YouTube video titled "Dialogue with B Madhava Puri, Michael Levin, Rasmus Haukedal, Bernd Rosslenbroich, and James Shapiro," an interesting and productive exchange between Dr. Rosslenbroich and Dr. Puri occurred where they distinguished between reduction and reductionism, and how to approach studying parts and wholes in an appropriate manner. In Dr. Rosslenbroich's aforementioned book Properties of Life: Towards a Theory of Organismic Biology, he distinguishes between reductionism and reduction. The former is the worldview that has dominated modern biology up until recent times, where wholes have no fundamental value and are merely aggregates of parts, while the latter recognizes that the whole is more than the sum of the parts and that these parts can only be understood in isolation up until a certain point, where eventually this highly specialized understanding must be reintegrated back into the whole along with all the other parts. Dr. Puri suggested that it is not enough to superficially acknowledge that a part belongs to a whole, and then immediately move on to studying the part in great detail. Rather, the relationship between the part and whole must first be thoroughly understood – even if it cannot be completely understood until after reduction occurs – before then moving on to narrowing focus onto the part to study exactly how it fulfills its purpose in relation to the whole. This subtle nuance, where parts can never become abstract independent entities but must retain their functional value in relation to the whole at all times, can make all the difference in understanding the truth about organisms and other living systems.

8. Processes of shape || This refers to the German word *Gestalt*, for which there is no sufficient English translation. *Gestalt* is the form of an organism, as well as its organs and tissues. It also refers to the process of developing such forms throughout an organism's lifetime. Attempts to reduce *Gestalt* to genetic information have been inconclusive. *Gestalt* should be scientifically studied at the appropriate level of biological organization (organism, cellular, molecular), while remaining cognizant of the interdependencies between these levels. Dr. Rosslenbroich argues that *Gestalt* is not an outdated science, as is sometimes claimed. Classical research areas in zoological morphology and paleontology have been overshadowed by reductionistic biochemical approaches, which Dr. Rosslenbroich feels is a misguided development. He believes that these areas still have significant contributions to offer to modern science.

9. Time autonomy || The sequential study of physiological processes – such as cell cycles, growth, maturation, heart rate, respiration, circadian rhythm, etc – reveals that they proceed in a very specific temporal order. This is time *autonomy* because these processes are self-generated time structures with unique self-generated durations. In the field of chronobiology, it is well established that these processes are not simply triggered by environmental factors. Time autonomy typically oscillates, meaning it is executed rhythmically. Dr. Rosslenbroich notes the dynamic nuances involved in interdependent layers of biological processes with varying oscillation frequencies – ranging from milliseconds, hours, a day, month, year, or longer – that overlap with each other. He elaborated on the circadian rhythm, where early research sought a central organ that triggered this rhythm, much like an internal clock. They identified a region in the anterior hypothalamus called the suprachiasmatic nucleus. This development led to a linear understanding of circadian rhythm, although more recent work shows the feedback loops in play among the interdependent players involved. Thus, the autonomous temporal processes are also integrated and irreducible to their constituents.

Due to time constraints on conference presentation times, the remaining properties discussed by Dr. Rosslenbroich were not as elaborately discussed. Please see his open-access book for more in depth analysis on these properties of life. The link to the book was provided at the beginning of this overview.

10. Sensitivity and affectability || The more developed an organism's autonomy through evolution, i.e., the more prominent its self-maintained independence from the environment is, the greater the capacity for exchange between an organism and the environment. Sensitivity is an organism's ability to perceive its internal states as well as certain conditions of its environment, while affectability refers to an organism's capacity to have its internal states impacted by the environment, beyond the effects of mere physical force of environmental stimuli.

11. Subjective experience || This begins at the level of single cells, where they have an experience of themselves as distinct from the environment (subject and object distinction). As living entities become increasingly complex, this property becomes more intense. Dr. Rosslenbroich suggests that this might be the basic principle which generates consciousness.

12. Ability to evolve || The world of organisms is in a constant long-term change. Dr. Rosslenbroich distinguished between simply describing this evolutionary history of organisms and actually understanding the factors driving the process. He states that, over the last two decades, it has become clear to scientists that random variation with natural selection is not sufficient to describe evolution. There are a wide range of other factors relevant to evolution that are being discussed, and Dr. Rosslenbroich feels that the organismic view of biology will significantly contribute to this extended evolutionary synthesis. Further material in this regard can be found on www.thethirdwayofevolution.com.

13. Growth and development || Dr. Rosslenbroich posed the question: Why is it necessary for organisms to constantly undergo processes of reconstruction (i.e. growth and development)? We've become used to this as a normal part of life, so we might take its necessity for granted, but this is a valid and important question. Factually, there is no such thing as a static organism, but rather the organism is constantly in a process of change or becoming. Reducing the totality of an organism to a single stage of its development, like adulthood for instance, is really just a snapshot of the truth of its living dynamic activity.

14. Environmental interrelationships || Today, more and more scientists are recognizing that the relationship between organisms and the environment is not one-way. Its not only that organisms adapt to the environment, but they also construct it. Dr. Rosslenbroich reiterates that he is just making small points about a big discussion due to the conference's time constraint.

15. Reproduction and death || Death processes are not the opposite of living ones; rather death is a part of life. This becomes especially clear when studying apoptosis, or programmed cell death, where, within organisms, there is regularly generataion and degeneration of cells. This principle is also presnet at the ecosystem level, where death occurs so that new life can flourish. Dr. Rosslenbroich quoted Johann Wolfgang von Goethe, who said "[Nature's] spectacle is always new, because she always creates a new audience. Life is her most beautiful invention, and death is her artifice to have abundant life."

The presentation concluded by acknowledging the challenge of starting to think in an integrative way that perceives the interconnection between these 15 distinct properties of life, without reducing them to physicochemical properties. This requires scientists to move away from seeking linear causal chains and start looking for nuanced integrated webs of interdependent interactions. Dr. Rosslenbroich feels that humanity's attitude towards ethical, medical, agricultural, and ecological considerations of life in general will change for the better once we start thinking in an organismic way, which starts by taking these 15 properties as serious starting points for research. Dr. Rosslenbroich closes by sharing that shifting humanity's mentality to think in terms of contexts, interdependencies, networks, multi-layered structures of autonomy, etc, we will also be moving closer to a spiritual way of life. This is due to the fact that these properties of life suggest that there is a high degree of order inherent in reality, where order or organization is not only material and mechanical, but indicative of purpose and intelligence. The more we can think about and reflect on this order, we can follow the inherent principles of organization within natural processes to better understand them. We are not strangers to the inner workings of reality, but part of it. Thus, our consciousness or intelligence, in addition to our capacities to connect with nature emotionally and empathically, are not external or foreign to reality, but are participatory within reality as a whole. This participation extends beyond the biological aspect of life to include our thoughts and feelings. Thinking within nature or reality as participants rather than alienated observers reveals a spiritual dimension to our experience, which can benefit our scientific investigations.

5

James A. Shapiro

Why Evolution Works: Life Doesn't Wait for Accidents — Life Changes Itself

James A. Shapiro is Professor of Microbiology emeritus in The Department Of Biochemistry And Molecular Biology at the University Of Chicago. He received his Ph.D. in Genetics from Cambridge University in 1968 under Prof. W. Hayes, FRS. At the University of Chicago since 1973, he was Darwin Prize Visiting Professor at the University of Edinburgh (1993). In 2001, he received an O.B.E. for services to the Marshall Scholarship Program. He is a founder of www.TheThirdWayofEvolution.com, intended to raise awareness of scientific alternatives to Intelligent Design and Neo-Darwinism. His pioneering books are on mobile genetic elements, natural genetic engineering, and bacterial multicellularity. His complete CV can be found at https://shapiro.bsd.uchicago.edu/cv.shtml.

Dr. James A Shapiro's presentation offered evidence for how life changes itself through evolution, demonstrating that cellular and genome modification occurs in evolution organically or biologically, rather than through changes resulting from random accidents. He defines evolution as "descent with modification." Before proceeding, the following clarification of terms may be helpful for some readers. Chromosomes, DNA, and proteins are fundamental components of genetics and cell function. DNA is the molecule that carries genetic instructions. Chromosomes are organized structures made of DNA and proteins that help package and manage the DNA. Proteins are the functional molecules built using the instructions encoded in DNA, and they perform most of the work in cells. We now know that non-coding ncRNAs also do significant biological work in living cells.

Due to random mutations possessing "vanishingly small" probabilities of producing significant changes to the genetic code, they cannot explain evolutionary variation. Further, more than 99% of so-called spontaneous mutations are derived from repeatedly observed biological activity, in that they're "enzymatically induced by mutator polymerases, cytidine deaminases, and other DNA-modifying enzymes." Dr. Shapiro explains that greater than or equal to "99.9% of incorporation errors are removed by replication proof-reading functions." Examples of these functions occur during replication when a DNA polymerase enzyme adds an incorrect nucleotide base, where accurate pairing matches adenine (A) and thymine (T) base pairs or guanine (G) and cytosine (C) complementary bases, and incorrect pairing matches A or T to G or C. In such cases, the polymerase detects the mismatch and uses exonuclease activity to remove the improper nucleotide from the end of the DNA molecule. After replication, if misincorporated nucleotides remain, a process called post-replication mismatch repair occurs. In this process, the new DNA strand is cut, the incorrect nucleotide and neighboring bases are removed, and the polymerase inserts a correct segment. Then, a DNA ligase seals the gap in the DNA backbone to complete this proof-reading function. Spontaneous changes and replication errors are kept to a very low level, thus their impact on cellular and genome modification is minimal. With these phenomena in mind, Dr. Shapiro goes on to describe events that require cleavage and ligation of extended DNA segments in evolution. Here, DNA cleavage is the splitting of DNA molecules by severing the bonds in its backbone, and DNA ligation is the joining together of two DNA fragments, both of which have been previously described. Dr. Shapiro refers to this cognitive functionality of cells as "natural genetic engineering," which is present in all living cells and is expressed through various systems.

The first event requiring cleavage and ligation mentioned by Dr. Shapiro was Horizontal Gene Transfer (HGT) observed in nematode worms and beetles, which have gained the ability to digest plant material. Plant material is typically challenging to digest, but bacteria and fungi possess enzymes that can break down such material. Through single-step acquisition by HGT, nematode worms and beetles acquire the genetic material necessary to build one of the enzymes that break down plant polymers from these microbial sources, thus gaining the ability to consume plant material. This is significant because, traditionally, biology has taught that organisms can only receive genetic material by inheritance from their parents. However, developments in 21st-century biology have disproved this. As Dr. Shapiro explained, DNA acquisition also includes HGT, which occurs across all taxonomic/species boundaries and involves (1) excising the DNA from the donor organism, (2) transferring it to the new recipient organism, and (3) integrating the acquired DNA into the recipient through natural genetic engineering.

Another event Dr. Shapiro describes is the rearrangement of protein domains. He explains that proteins are not unitary molecules. Individual proteins are composed of sections called domains, which have unique structures, activities, and interactions. The same domain can be found in different proteins. Dr. Shapiro states that protein domain rearrangements, i.e., acquiring or losing DNA encoding specific domains, result in combinatorial functional changes that occur much faster than single-nucleotide modifications (postulated in conventional accounts) in altering the activity of proteins. Changes in encoded proteins also occur after viral endogenization, a process in which genetic material from viruses integrates into a host's genome. An example is the role of endogenous retroviruses in the evolution of the placenta. Dr. Shapiro acknowledges the significant difference between viral genetic coding sequences and cellular sequences, and admits that scientists still don't quite understand the origin of viral sequences. Major hereditary changes can also happen after the rewiring of transcriptional networks by transposable elements (transposons). Transposons — sections of DNA which are capable of moving from one place in the genome to another — were discovered by Nobel laureate Barbara McClintock in the 1940s, of whom Dr. Shapiro was a close colleague. When they move, transposons carry control sequences (DNA sequences that serve as binding sites for transcription factors) that are then inserted into different locations throughout the genome. Transcription factors are molecular switches regulating when and how genes are turned on or off in a cell. When two or more transposons are inserted into different coding regions, these regions can become coordinated and respond to the same control sequences and transcription factors. This helps build the regulatory networks necessary for organismal development, where network development, rather than simple gene changes, accounts for the majority of developmental processes. Dr. Shapiro offers examples such as AmnSine-MER117's role in frontonasal development, L2's role in mammary gland evolution, and MER418's role in innate immunity.

Dr. Shapiro presented a series of figures illustrating the principles from his explanations of events that require the cleavage and ligation of extended DNA segments in evolution. The first figure was from the original human genome sequencing publication in 2001, showing examples of domain accretion (the formation of multidomain proteins) in chromatin proteins. This depicted three instances of how specific proteins change by acquiring new groups of extended domain regions. Another figure illustrated the expansion of transcription factor domains and architectures, showing ancient architectures shared by all animals, architectures exclusive to flies and humans, ones unique to flies and worms, and finally, domains unique to human beings. Sharing these two figures was intended to reinforce the revolution in scientists' understanding of how protein structure and evolution occur. The third figure showed some of the vast array of transposable elements that have been identified, including DNA Transposons, Retrotransposons, Long Interspersed Nuclear Elements (LINEs), and Short Interspersed Nuclear Elements (SINEs). LINE transposable elements constitute 21% of the genome, while DNA coding proteins is 1.5% of the genome. Notably, this demonstrates the genome's flexibility. Dr. Shapiro added that "all of the coding regions of the genome are, in principle, transposable." All of this experimental evidence challenges the notion of simple genetic determinism, the belief that protein-coding genes are the sole determinants of an organism's physical and behavioral traits.

Offering another visual to drive these points home, Dr. Shapiro discussed the peppered moth. The peppered moth, often regarded as a poster child for conventional evolutionary theory, ironically highlights the importance of transposable elements in shaping phenotypes, thereby challenging traditional evolutionary theory. Initially, this moth had an off-white color with black specs scattered across its body and wings. As the Industrial Revolution progressed, buildings and trees became covered by black soot, which made these whitish moths stand out, endangering their survival. Over time, the peppered moth became all black. Biologists initially thought this transformation was due to repeated random mutations and natural selection; however, molecular biologists later discovered that the insertion of DNA transposons was responsible for the newly formed black body color.

The third and final set of events requiring cleavage and ligation mentioned by Dr. Shapiro were chromosome rearrangements, where double-strand breaks in the genome are corrected by end-to-end joining of broken ends by alternative repair systems. Focusing on the consequences of chromosome restructuring, chromosome restructuring requires the fusing of broken ends of two chromosomes or ends of the same chromosome broken at different positions. This phenomenon was described in a 2011 discovery that revealed cancer cells and germ line cells can generate novel genome sequences and chromosome structures at multiple sites simultaneously during a single cell division. Some of these findings were verified through direct examination in cell culture experiments, while others were confirmed only through genome sequencing. Dr. Shapiro shared a visual of four circle plots of the human genome in cancer patients. These showed chromosome rearrangements resulting from chromothripsis (shattering of a chromosome), where rearrangements are generated by haphazard joining of the shattered fragments. He also shared a video of chromothripsis. This showed a cell culture experiment where chromosomes within the nucleus of a cell were undergoing DNA replication (colored green) while a micronucleus (colored red) containing a fragment of one of those chromosomes (created by being cut by the CRISPR genome editing system) shatters into dozens of smaller pieces.

Repairing chromosome breaks produces chimeric genome sequences that can produce novel products, where the protein domains from the two different joined chromosomes (thus it's a chimera) are connected and produce something new. It can also generate different regulatory patterns of genome expression. Dr. Shapiro called attention to the bizarre but well-documented phenomena of DNA polymerase "template switching" during the repair process to produce multiple chimeric sequences. These multisite chromosome rearrangements occur by the repair DNA polymerase taking parts from other chromosomes as it executes end joining. This can create chimeras within chimeras which equates to a virtually unlimited capacity to generate new genetic sequences. Another feature of rearrangements is that chromosomes spatially close to each other affect the expression of sequences of nearby proteins. If they're rearranged (thus altering their regulatory interactions), a different program of gene expression is generated, which produces a different phenotype that may be useful for exploiting the environment more effectively. Restructured chromosomes do not pair in the same way as their normal parent chromosomes, and can inhibit sexual reproduction by hindering gamete formation (the generation of sperm and egg cells).

Dr. Shapiro explains that these changes relate to speciation (the transformation of one species into another) in the following way. Hybrid speciation is one way to create new species quickly. Here, two different species that are reproductively compatible produce offspring whose germline cells' genomes are unstable and experience rapid change in chromosome structure and activation of transposable elements. If properly bred, such hybrid offspring can belong to a new species that is reproductively isolated from both parent species within two or three generations. This hybrid speciation has been demonstrated in *Heliconius* butterflies, of which Dr. Shapiro shared a graphic showing the common phenotypes. Other examples include *Saccharomyces* and *Candida* yeast, plant parasitic nematodes, Swallowtail butterflies, East African chiclid fishes, Darwin's finches (Galapagos), sparrows, clymene dolphins, flowering plants like sunflowers, orchids, and irises, and finally many crops such as emmer wheat, flour wheat, rapeseed, tobacco, peas, coffee, potatoes, cotton, rice, soybeans, and quinoa.

Dr. Shapiro's presentation started to close by discussing the role of biocognition in evolutionary changes. Cognitive processes regulate the biological functions underpinning transformations in the genome that give rise to evolutionary change. Cognitive responses are non-random responses. Biocognitive processes *initiating* evolutionary changes include:

- Quorum sensing pheromones enable bacteria to determine their population density by triggering the uptake of DNA and facilitating direct cell-to-cell DNA transfer. This phenomenon is particularly pronounced in biofilms, where most bacteria reside.

- In yeast, mating pheromones trigger oriented growth and cell fusions.

- Transposable elements are activated in all cell types by abiotic stresses such as wounding, radiation, heat, toxic metals, peroxide and other Reactive Oxygen Species, insecticides, and other synthetic organic chemicals.

- Genetic recombination (recombinases), transposable elements, targeted DNA cleavages, and changes in ploidy (the number of complete sets of chromosomes in a cell such as haploid, diploid, etc) in many cell types are activated by biotic stresses such as virus infections, toxins, predators, DNA transfection (artificially introducing DNA or RNA into cells), telomere erosion, replicative stress, inflammation, chromosome Breakage-Fusion-Bridge cycle (discovered by Barbara McClintock), and tissue culture.

- Transposable elements (transposons) and retrotransposons in bacteria, yeast, and mice are activated by starvation.

- Epigenetic regulation, transposon activation, stimulation of chromosome rearrangement, and initiation of whole-genome duplication and rapid speciation in eukaryotes ranging from yeast, flowering plants, insects, and vertebrates — these are all triggered by interspecific hybridization.

Biocognitive processes *executing* evolutionary changes include:

- Coordinating one cell engulfing another during symbiogenetic cell fusions (the process involved in creating the mitochondrion in an ancestor of all eukaryotes), fertilization of egg cells by male gametes to generate embryos in sexually reproducing organisms, and the coordination of a pathogen infecting a cell during pathogenesis.

- Establishing mating pairs for prokaryotic horizontal transfers and eukaryotic sexual reproduction.
- The formation of multimolecular nucleoprotein complexes that are necessary for genome restructuring, such as DNA transposition, retrovirus retrotransposition, chromothripsis, and chromoplexy, which are all non-random events that need to be spatially organized in reference to the genome.
- The formation of intranuclear repair centers and the recruitment of damaged chromosomes to generate rearrangements by double-strand break repair systems. For example, when a yeast cell is injured from radiation exposure, it forms an intranuclear repair center where multiple damaged chromosomes travel to be fixed. This is clearly a cognitive process because a repair center for damaged chromosomes can't be built without knowing where those chromosomes are and where the damage is.
- Recognizing the lesions in DNA damage that require repair.
- Responding to the altered epigenetic status of interspecific hybrids.

Becoming aware of the biocognitive processes that initiate and execute evolutionary changes leads to a question that was forbidden to ask when science was bound to the deterministic presumptions of 20th-century biology: could cells have tools to *guide* evolutionary change adaptively in response to cognitive inputs? Dr. Shapiro concluded his presentation by offering three points containing suggestions for how cognitive cells might guide evolutionary change, which point to evidence previously discussed that is useful for answering such a question.

1. Stress and cell errors (like producing a micronucleus alongside the main nucleus during cell division) can activate genomic changes, both by stimulating transposable elements to move into

new locations in the genome, and also by responding to stress that leads to major DNA rearrangements (chromothripsis and other forms of chromoanagenesis).

2.   Transposable elements or transposons are influenced by proteins and DNA conditions, and don't just land anywhere. Transposons do not insert themselves randomly in the genome. Rather, they're usually directed to specific places by certain proteins. These places can have particular roles or be in a certain state, like being on or off. Transposons also tend to prefer certain kinds of genetic environments, like those with specific chemical tags (epigenetic marks) or activity levels (transcription). Additionally, transcription factors — proteins that help turn genes on or off, i.e., entities that influence genome activity levels — can help guide where transposons insert themselves.

3.   New traits often come from "genomic islands," chunks of DNA with groups of related genes. In eukaryotes, like plants and animals, these islands are linked to adaptive changes that help drive the formation of new species. In prokaryotes, such as bacteria, useful traits often come from horizontal gene transfer between organisms. Clusters of genes in certain DNA regions called integrons can lead to major new traits, helping species adapt or evolve, as happened in the case of multiple antibiotic resistance.

In summary, to experimentally validate that random genetic mutation has almost no role in biological evolution, Dr. Shapiro first explained how DNA proof-reading removes most errors caused by mutations. Then, he proceeded to a robust discussion of various systems of natural genetic engineering, where surprising transformations of DNA sequences are caused through biological functions that have just begun to be understood through progress in 21st-century biology. These fascinating systems disprove the 20th century's view of genetic determinism, where a fixed and rigid genome completely determined the form, function, and behavior of whole organisms. Examples of this biological activity that 21st-century biology is discovering, where DNA is cut (cleavage) and joined (ligation) through different systems, include: (1) Horizontal Gene Transfer (HGT), where organisms acquire DNA from other organisms of various species in the shared environment, rather than just through inheritance from parents. (2) Protein domain rearrangements, which give rise to significant genetic change faster than HGT due to a single domain being present in more than one protein molecule, thus a change/rearrangement of one domain leads to transformation of several proteins. Here, the role of Barbara McClintock's transposons triggering protein domain rearrangements was also discussed. (3) Chromosome rearrangements and their consequences, focusing on startling phenomena such as chromothripsis and template switching. After offering this impressive foundation of experimental evidence from molecular biology, showing how dynamic and biologically-influenced the genome truly is, Dr. Shapiro related this to organism-level speciation. He explained that reproduction between two different yet reproductively compatible species leads to progeny whose genomes are unstable and undergo the kinds of transformations previously presented. This can lead to new species that are reproductively incompatible with both parent species in two or three generations. Dr. Shapiro closed his presentation — which was focused on how evolution relies on living entities changing themselves through biological processes rather than requiring random accidents — by reframing the thorough experimental evidence offered throughout the talk around affirming a biologically progressive question: could cells have tools to *guide* evolutionary change adaptively in response to cognitive inputs?

During the Q&A following the presentation, organicist and evolutionary biologist Dr. Bernd Rosslenbroich inquired if it is plausible that organisms as a whole can trigger genetic changes. This question is a natural byproduct of Dr. Shapiro's talk, which presented how cells within the organism directly affect genetic changes. Dr. Shapiro's reply was, "I think we have to entertain the possibility, and we have to go about trying to find the evidence for it. I think that this can be done experimentally by taking interspecific hybrids, for example, and subjecting them to different inputs and observing if changes occur with any reliability. Nowadays, with the ease of genome sequencing, we should be able to do that with experimental organisms pretty readily." Then, philosopher Dr. Rasmus Haukedal asked about why, despite the overwhelming evidence offered in the presentation that has been established for decades, someone like Richard Dawkins can still promote a misleading gene's-eye-view of evolution. Dr. Shapiro pointed out that Dawkins has made a lot of money selling books with his version of things, and Dawkins may fear that going back on that might threaten his future earnings. Dr. Shapiro further responded that his colleague, Denis Noble, debated Dawkins in 2022 and presented some of the findings that Dr. Shapiro shared during this Science and Scientist 2024 conference, although Dawkins denied the importance of this work. Dr. Shapiro expressed his keen awareness of the issue of mainstream biologists not acknowledging evidence contrary to the dominant paradigm and shared that he even published a paper on "What prevents mainstream evolutionists teaching the whole truth about how genomes evolve?" (2021) in *Progress in Biophysics and Molecular Biology*. He concluded by saying that changes to how evolution is taught in schools will most likely not change until the general populace is more accepting of the conclusions of 21st-century biology, which show that evolution is a result of cognitive biological processes. This transition process is indeed underway. Further details are available at www.thethirdwayofevolution.com.

6

Interdisciplinary Dialogue

Proposing a New Scientific Methodology: Dr. Puri's Parting Gift to the Science & Scientist Conference Series

During the interdisciplinary dialogue at the end of the conference, the following question was raised: What experiments can distinguish between (a) the organismic level of organization exerting top-down agency on the cellular level and (b) the cellular level exerting bottom-up agency on the organismic level? This led to a very dynamic discussion among all the speakers that considered how different kinds of unities within an organism – from cellular networks, tissues and organs, to the organism as a whole – demonstrate various degrees of cognitive activity in pursuance of specific goals unique to their functions. Interestingly, everyone agreed that reductionism is untenable for progress in 21st-century biology.

Dr. Levin described his lab's approach to addressing this question: (1) make hypotheses about specific activity of the higher level of biological organization, (2) translate these hypotheses into a computational framework to derive implications of the proposed activity, (3) test the hypotheses through experiments. Testing the degree of intelligence at any given level of organization requires placing obstacles between the cognitive entity and its goal to determine whether and how it overcomes these challenges to reach the goal in unexpected ways. Following this response, Krishna Keshava Das, the panel discussion moderator, asked Dr. Levin how the diversified experience of cellular collectives scales up to a unified intelligence of the organism as a whole with a singular identity, as experienced first-person by most waking people. Dr. Levin echoed the doubts he had already expressed in his presentation that there is such a thing as a "true unified intelligence." The authors of this conference proceedings have elaborated on this in the section focused on Dr. Levin's talk, especially in the subsection titled "***The Reality Of Diversity Does Not Preclude The Truth Of Unity***," where we offer challenges.

Das further inquired about the intuitive notion that, as a human organism with a first-person experience of being a unified self-identity, such experience seems to have causal or agential priority over the sub-unities in the body, i.e., organs (unities of tissues and cells). He argued that if an organ's cognitive activity disturbs the organism's activity, where this would only occur in a diseased state of the organ, then the organ is removed from the body and potentially replaced with a healthy one. Organs, on the other hand, cannot reject the organism as a whole, indicating a hierarchy. Dr. Levin denied that the organism's cognitive experience has priority over that of its organ systems. He said that a person believes that their first-person experience takes precedence over the individual experiences of their kidneys, brain, liver, etc., because the left hemisphere of the brain is linguistically capable, whereas most other organs are not. While Dr. Levin's work depends on this kind of skepticism in order to discover what kinds of hidden intelligence varying levels of biological organization possess, which his lab excels in, his dismissal of the priority of the whole in relation to the parts became a point of contention with some of the other discussion participants.

From this point forward, the dialogue evolved into a discussion of the whole-part ontology that modern science tends to presuppose, the epistemological method by which this ontology is determined and applied in experiments, and a proposal for a new onto-epistemological approach science.

Dr. Puri entered the dialogue by cautioning against mistaking scientific models for the actual phenomena, a logical fallacy sometimes referred to as mistaking the map for the territory. This is another kind of reductionism. He directly addressed Dr. Levin's point about denying a hierarchy between the cognitive life of the organism as a whole and its organs, and introduced the Vedic conception of *Bija* to indicate that diversity arises from unity. Here, unity does not imply homogeneity, but rather heterogeneity. Ancient Vedic wisdom describes the entire universe as arising from a seed (*Bija*), the primordial unity that differentiated into natural kinds throughout material creation. Each living entity also arises from a *Bija*. The unitary seed implicitly contains the totality of the whole, and the seed's unfolding development gives rise to different parts that are functionally related to and interdependent on the whole. This is observed throughout the kingdom of life. In a mature oak tree, leaves, branches, the trunk, and roots all manifest from a singular seed. Dr. Puri emphasized the observation that diversity arises from unity, not that unity is imposed on diversity after it has developed, indicating a hierarchical dynamic at work.

Drawing on his professional background as a quantum chemist, Dr. Puri reinforced the above point from a quantum physics perspective. He explained that, in quantum mechanics, the whole or totality of reality is called the square of the wave function (the wave function multiplied by itself). Then, an operator acts on the square of a wave function to produce an observable. The totality is there from the beginning, not contingently constructed from observables. The observables are derived from the preexisting totality. With this in mind, Dr. Puri explained that reductionism arises from treating immediate observations as the truth, without recognizing that observation limits the totality to a single entity or set of entities.

Further demonstrating the validity of the *Bija* conception, Dr. Puri referred to Dr. Shapiro's presentation title, "Life Changes Itself." This idea implies an inner division of life in which some aspects change while a unitary identity endures through transformation. Dr. Puri reiterated that life is not an accumulation, but a division of itself, like a seed developing to maturity. Embryogenesis is the process by which the germ or seed produces new molecules that enable it to determine itself and mature. The cognitive entities within the body do not compete with the organism as a whole; rather, they develop for it and are integrated into it.

At the end of his exchange with Dr. Levin, Dr. Puri argued that accepting abiogenesis as the origin of life, the notion that a unitary living entity arose from a diversity of preexisting molecules, is fundamentally reductionistic in that it presumes unity came from diversity, and is also completely in the realm of abstract thought since such an event has never been observed in nature nor produced in a scientific lab. Dr. Puri explained that our thoughts about nature must be grounded in observation, and that the *Bija* concept meets this requirement, whereas abiogenesis does not. Before departing, Dr. Levin encouraged Dr. Puri to apply the *Bija* model to experimental development aimed at scientific discovery. Dr. Levin said that if it can produce more tangible, positive benefits than the existing abiogenic model, he would be willing to accept it. We hope that these conference proceedings and the Princeton Bhakti Vedanta Institute's future work contribute to this worthy goal.

A rich exchange continued between Dr. Puri, Dr. Rosslenbroich, and Dr. Haukedal. Picking up on Dr. Puri's quotation of Hegel that "the truth is the whole," Dr. Haukedal sought to clarify that Hegel did not espouse a solely top-down perspective with emphasis on the causal power of the whole alone. Dr. Puri agreed and explained that the whole determining the parts is only half the story; the parts also determine the whole. Hegel offered a dialectical view, similar to Immanuel Kant's understanding of an organism, in which both sides are causally interrelated. This is a circular view of causality, where a cause does not just linearly precede an effect, but that cause was the effect of a previous cause, and the effect can be the cause of something else. In this regard, Dr. Puri appreciated Dr. Rosslenbroich's presentation, which offered a model of circular causality in biological activity. While Dr. Rosslenbroich agreed with everything Dr. Puri had said, he reminded the group that the distinction between reduction and reductionism remains important, so that a scientific methodology that values detailed knowledge of the parts is retained. While reductionism makes an unjustified ontological claim about wholes being nothing more than the sum of their parts, reduction acknowledges the fundamentally interdependent whole-part relation and seeks to bring knowledge of the part into an integrative understanding of the whole. This was elaborated in Dr. Rosslenbroich's presentation. While Dr. Puri appreciated this idea and the value of parts, he also pushed back against the methodology of reduction, as it only ambiguously accounts for the whole while rushing to analyze parts, which he pointed out are never ending. Dr. Puri insisted that knowledge of the whole must be the starting point, since the parts cannot be understood truthfully outside their relation to the whole, and emphasized that the overarching connection among the parts cannot be determined after the parts are understood. The nuance distinguishing Dr. Puri's perspective can be quite subtle, but it makes all the difference. By starting with the whole and its purposes, identifying its functional parts, studying the specific function of each part in relation to the whole, and then examining the individual part in greater detail to ascertain how it executes that relational function, we start with the unity and trace the diversity in its connection to the overarching totality. This allows us to see the role of the whole in the part, which is overlooked by both methodologies of reductionism and reduction. When asked about his thoughts on the discussion, Dr. Shapiro stated, "I think that this kind of discussion is very rich, and it makes me appreciate what we're trying to do as thinkers and scientists. How we put the parts and the whole back together again is the secret of how we understand what is going on."

Dr. Puri elaborated that his proposed methodology entails considering the whole as psychophysical. It is not merely a physical thing but also a thought. To demonstrate this, he explained that scientists first hypothesized the existence of an electron, and only after positing it did they observe a phenomenon in nature that corresponded to their theory. Thus, Einstein said, "Whether you can observe a thing or not depends on the theory which you use. It is the theory which decides what can be observed." That is why Hegel said everything must be understood as both subject and substance. Due to the history of Western modern science turning a blind eye to the role that the consciousness of the scientist plays in their observations of and conclusions about nature, the present scientific method presupposes the epistemology of empiricism as the means for arriving at true conclusions about natural phenomena. But empiricism only reveals an immediate diversity of sensory data; it doesn't provide access to the holistic context that initially gives rise to and ultimately unifies such data. Thus, a scientific method that presupposes empiricism leads to a reductionist view of nature, resulting in a mechanistic account of life. In Dr. Puri's book *Idols of the Mind vs. True Reality* (2020), he used the ancient Indic blind men and the elephant parable to illustrate the downfall of attempting to construct a whole by analyzing the parts in isolation from the whole. So the immediacy of empiricism is at the root of reductionism. Observation limits the whole to a set of its parts within the purview of immediate sense perception. The totality remains unknown in the nuances of consciousness, which is presupposed by empirical observation. Starting with empiricism is synchronic, reducing reality to the immediate existent (material and efficient aspects of Aristotelian causality). Starting with the conceptual (the formal and final aspects of Aristotelian causality) is diachronic, seeing holistically that consciousness precedes and contextualizes existence. The formal aspect of cause is the seed, and the final aspect is the mature functional relationality between part and whole. Here is a real framing for systems thinking.

Progress beyond this panel discussion hopefully includes determining how scientists from various disciplines can learn to think about the whole first, then identify its functional parts, ascertain the details of the functional relations, and finally reduce to how individual parts execute the function, thereby offering deeper insight into the whole itself and its role in the part. Perhaps this deeper view of the whole, which results from following this method, will trigger another round of following the method to discover ever-new functional aspects and individual expressions, leading to ever-deepening meditation on nature and its truth as Spirit. This is a call for a new scientific methodology that begins with consciousness and self-reflection, then proceeds to empiricism within this context, rather than starting with empiricism as in the current established scientific method.

A progressive scientific methodology should acknowledge the dialectic between ontology (the study of the known) and epistemology (the study of how we know). This reciprocal dynamic between how reality is and how we think about reality gives rise to the necessity of theology (the study of the knower). Our scientific method must align our inquisitiveness and conception of reality with reality as it is; this is the fulfillment the knower seeks by pursuing true knowledge. Dr. Puri suggested that Hegel's project has already, to a considerable extent, provided the world with a framework for this ontoepistemic approach to science. Hegel's own philosophical system is a result of this method. Hegel starts with what is most simple or presuppositionless, and demonstrates its necessary conceptual development to the most highly determined, which is connected to the starting point as a circle or "circle of circles." That presuppositionless beginning from which all else unfolds is not starting with the part to scale up to knowledge of the whole, but rather is starting with the seed or *Bija* within which the whole unity-in-diversity exists implicitly, in order to allow its conceptual development to unfold and make everything explicit.

Das, K. K. (2025). "A 21st-Century Environmental Ethic: Theistically-Conscious Biocentric and Biomimetic Innovation." *Religions* 2025, 16(8), 1038; https://doi.org/10.3390/rel16081038.

Das, K. K. (2023). "Towards a Science of Spirit – A Holistic Approach to Big Questions in Modern Science." *Global Conversations: An International Journal in Contemporary Philosophy and Culture*, Vol. 6., pp 80-104.

Muni, B. V. (2019). "Quantum Mechanics Shows the Limit of Naïve Realism" in *Quantum Reality and Theory of Śūnya*, 249-269. Siddheshwar Rameshwar Bhatt (ed.). Springer Singapore. doi.org/10.1007/978-981-13-1957-0

Shanta, B. N. (2019). "Subjective Evolution of Consciousness in Modern Science and Vedāntic Philosophy: Particulate Concept to Quantum Mechanics in Modern Science and Śūnyavāda to Acintya Bhedābheda-Tattva in Vedānta" in *Quantum Reality and Theory of Śūnya*, 271-282. Siddheshwar Rameshwar Bhatt (ed.). Springer Singapore. doi.org/10.1007/978-981-13-1957-0

Shanta, B. N. (2018). "21st century biology establishes that life is beyond chemicals." *International Journal of Recent Trends in Science And Technology*, P-ISSN 2277-2812 E-ISSN 2249-8109 Special Issue, ACAEE: 2018 pp 311-329

Shanta, B. N., Muni, B. V. (2016). "Why Biology is Beyond Physical Sciences?" *Advances in Life Sciences*, 6 (1): 13-30. doi:10.5923/j.als.20160601.03

Shanta, B. N. (2016). "Vedāntic view of life: Reply to Gustavo Caetano-Anollés." *Communicative & Integrative Biology*, 9(2), e1160191. doi.org/10.1080/19420889.2016.1160191.

Shanta, B. N. (2015). "Life and consciousness – The Vedāntic view." *Communicative & Integrative Biology*, 8(5), e1085138. doi.org/10.1080/19420889.2015.1085138.

Shanta, B. N. (2014). "The Chronology of Geological Column: An Incomplete Tool to Search Georesources." In *Geo-Resources*, 609-625. K. L. Shrivastava, Arun Kumar, P. K. Srivastav, and H. P. Srivastava. Jodhpur, Rajasthan, India: Scientific Publishers. doi:10.13140/RG.2.1.4409.4808

The Bhakti Vedanta Institute of Spiritual Culture and Science (BVISCS) is a 501(c)(3) charitable nonprofit educational organization located in Princeton, New Jersey, USA. Our mission is to promote a more compassionate attitude towards the environment and nurture the development of the spiritual dimension of life in education, science, and society.

Since its inception in 1974, the Bhaktivedanta Institute (BI) has drawn upon insights from Bhagavat Vedanta, the distilled essence of the vast ancient Vedic wisdom, to address problems in modern science related to the reduction of life to matter, cosmological models, and the role of consciousness in empirical investigations of nature. Today, the BI comprises numerous autonomous institutes worldwide that engage in interdisciplinary studies at the intersection of science, philosophy, and religion.

The Princeton BVISCS, **established in 2012** by BI founding charter member Bhakti Madhava Puri, PhD, uniquely derives insights from Hegelian philosophy, which complements the Bhagavat as affirmed by stalwart Vaishnava Acharyas (Spiritual Masters). For the first eight years of operation, BVISCS was situated right across the street from Princeton University. Still based in Princeton, the BVISCS continues **the work that Dr. Puri has been dedicated to for over 50 years.** Informed by the history and philosophy of science, servitor scholars from the BVISCS address questions concerning the origin, purpose, and evolution of matter, life, consciousness, and self through the lens of Organic Wholism. The BVISCS strives to foster cultural and scientific progress beyond materialistic values and medical/technological advancement. Humans live longer and have fancier gadgets, yet suffer worldwide from an environmental and mental health crisis. We don't know how to harmonize with our surroundings, and our minds have grown uneasy. Through **publications** and the **Science & Scientist conferences**, the BVISCS's annual international conference series, we discuss Dr. Puri's progressive work with other thinkers who are themselves engaged in relevant investigations across the sciences and humanities. If you discover a mutual interest in this work, or would like to attend our Bhagavat cultural programming, kindly consider **reaching out** (princeton@bviscs.org).